AF566667

Functional Morphologic Changes in Female Sex Organs Induced by Exogenous Hormones

Edited by
Gisela Dallenbach-Hellweg

With 139 Figures and 42 Tables

Springer-Verlag
Berlin Heidelberg New York 1980

Professor Dr. *Gisela Dallenbach-Hellweg*
Universitäts-Frauenklinik der Univ. Heidelberg
Klinikum Mannheim, Morphologische Abteilung
Postfach 23, 6800 Mannheim, West-Germany

ISBN-13:978-3-642-67570-6 e-ISBN-13:978-3-642-67568-3
DOI: 10.1007/978-3-642-67568-3

Library of Congress Cataloging in Publication Data.
Main entry under title:
Functional morphologic changes in female sex organs induced by exogenous hormones.
Bibliography: p. Includes index.
1. Generative organs, Female – Drug effects. 2. Estrogen – Physiological effect. 3. Progestational hormones – Physiological effect.
I. Dallenbach-Hellweg, Gisela. [DNLM: 1. Genitalia, Female – Cytology. 2. Genitalia, Female – Drug effects. 3. Sex hormones – Pharmacodynamics. WP100 F979]
QP265.F86 615'.766 80-10863
ISBN-13:978-3-642-67570-6

Softcover reprint of the hardcover 1st edition 1980

Composition: SatzStudio Pfeifer, Germering.

2121/3321-543210

Contents

II. Gestagens

III. Combined Hormones

List of Contributors

Bässler, Prof. Dr. R.
Pathologisches Institut der Städt. Kliniken Fulda, Pacelliallee 4, 6400 Fulda, West-Germany

Baulieu, Prof. Dr. E.-E.
INSERM U 33, Faculté de Médecine, Université Paris-Sud, Lab. Hormones, F-94270 Bicêtre, France

Beyer, Dr. Johanna
Endokrinologische Abteilung, Univ.-Frauenklinik, Klinikum der Albert-Ludwigs-Universität, Hugstetter Straße 55, 7800 Freiburg, West-Germany

Bonk, Dr. U.
Institut für Pathologie, Zentralkrankenhaus Bremen-Nord, Hammersbecker Str. 228, 2820 Bremen 70, West-Germany

Clocuh, Dr. Y.P.
Institut für Klinische Zytologie der Technischen Universität (Klinikum rechts der Isar), Prinzregentenplatz 14, 8000 München 80, West-Germany

McClure, M.D., H.M.
Yerkes Regional Primate Research Center, Emory University, Atlanta, Georgia 30322, USA

Collins, M.D., D.C.
Department of Medicine, Emory School of Medicine, Atlanta, Georgia 30303, USA

Czernobilsky, Prof. M.D., B.
Department of Pathology, Kaplan Hospital, Rehovot, Israel

Dallenbach, Prof. Dr. F.D.
Institut für experimentelle Pathologie, Deutsches Krebsforschungszentrum, Im Neuenheimer Feld 280, 6900 Heidelberg, West-Germany

Dallenbach-Hellweg, Prof. Dr. Gisela
Morphologische Abteilung, Univ.-Frauenklinik, Klinikum Mannheim, Postfach 23, 6800 Mannheim, West-Germany

Diczfalusy, Prof. Dr. E.
Reproductive Endocrinology Research Unit, Swedish Medical Research Council and Dept. of Obstetrics and Gynecology, Karolinska sjukhuset, Stockholm, Sweden

Egger, Priv.-Doz. Dr. H.
Frauenklinik mit Poliklinik und Hebammenschule der Universität Erlangen-Nürnberg, Universitätsstraße 21/23, 8520 Erlangen, West-Germany

Ferenczy, M.D., A.
Associate Professor of Pathology and Obstetrics and Gynecology of the Sir Mortimer B. Davis Jewish General Hospital and McGill University, 3755 Cote Ste. Catherine, Montreal, PQ H3T 1E2, Canada

Graham, Ph.D., C.E.
Deputy Director, International Center of Environmental Safety, Albany Medical College, P.O. Box 1027, Holloman Air Force Base, New Mexico 88330, USA

Grumbrecht, Prof. Dr. C.
Geburtshilflich-Gynäkologische Abteilung des Diakonissenkrankenhauses, 7500 Karlsruhe-Rüppur, West-Germany

Herbst, Prof. Dr. R.
Pathologisches Institut der Freien Universität Berlin, Universitätsklinikum Charlottenburg (FB 3), Spandauer Damm 130, 1000 Berlin 19, West-Germany

Johannisson, Prof. Dr. Elisabeth
Centre de Cytologie et de Déspitage du Cancer, Boulevard de la Cluse 51, 1205 Geneve, Switzerland

Joswig-Priewe, Dr. H.
Pathologisches Institut des Städt. Krankenhauses, Robert-Koch-Straße 1, 8300 Landshut, West-Germany

Kindermann, Prof. Dr. G.
Frauenklinik u. -Poliklinik (WE 14), Freie Universität Berlin, Klinikum Charlottenburg (FB 3), Pulsstraße 4–14, 1000 Berlin 19, West-Germany

Kranzfelder, Dr. D.
Universitäts-Frauenklinik und Hebammenschule Würzburg, Josef-Schneider-Str. 4, 8700 Würzburg, West-Germany

Kubatsch, Dr. B.
Pathologisches Institut der Universität des Saarlandes, 6650 Homburg, West-Germany

Kühnel, Prof. Dr. W.
Abt. Anatomie der Med. Fakultät an der Rhein.-Westf. Techn. Hochschule Aachen, Melatener Straße 211, 5100 Aachen, West-Germany

Kurman, M.D., R.J.
Associate Professor of Pathology and Obstetrics and Gynecology, Georgetown University, School of Medicine, Washington D.C., USA

Laakso Dr. K.
Praxis für Pathologie, Postfach 1469, 8130 Starnberg, West-Germany

Landgren, M.D., Britt-Marie
Reproductive Endocrinology Research Unit, Swedish Medical Research Council and Dept. of Obstetrics and Gynecology, Karolinska sjukhuset, Stockholm, Sweden

Maass, Dr. H.
Univ.- Frauenklinik und -Poliklinik Hamburg-Eppendorf, Martinistraße 52, 2000 Hamburg 20, West-Germany

Masironi, M.D., Britt
Reproductive Endocrinology Research Unit, Swedish Medical Research Council and Dept. of Obstetrics and Gynecology, Karolinska sjukhuset, Stockholm, Sweden

Mestwerdt, Priv.-Doz. Dr. W.
Universitäts-Frauenklinik und Hebammenschule Würzburg, Josef-Schneider-Str. 4, 8700 Würzburg, West-Germany

Mortel, M.D., R.
INSERM U 33, Faculté de Medécine, Université Paris-Sud, Lab. Hormones, 94270 Bicêtre, France

Multier-Lajous, Dr. A.-M.
Institut für Pathologie der Freien Universität Berlin, Universitätsklinikum Charlottenburg (FB 3), Spandauer Damm 130, 1000 Berlin 19, West-Germany

Nilsson, Prof. Dr. B.O.
Reproduction Research Unit, Biomedical Centre, P.O. Box 571, 751 23 Uppsala, Sweden

Norris, Prof. M.D., H.J.
Department of Gynecologic and Breast Pathology, Armed Forces Institute of Pathology, Washington D.C. 20306, USA

Pape, Dr. C.
Univ.-Frauenklinik und -Poliklinik Hamburg-Eppendorf, Martinistraße 52, 2000 Hamburg 20, West-Germany

Pöschl, Dr. U.
Praxis für Pathologie, Postfach 1469, 8130 Starnberg, West-Germany

Prechtel, Prof. Dr. K.
Praxis für Pathologie, Postfach 1469, 8130 Starnberg, West-Germany

Schaude, Dr. H.
Geburtshilflich-gynäkologische Abteilung des Kreiskrankenhauses, Friedhofstraße 13, 7967 Bad Waldsee, West-Germany

von Schilling, Dr. B.
Merck Quimica, S.A., Laboratorio de Bio-Investigación, Apartado 47, Hospitalet de Llobregat, Barcelona, Spain

Schirmacher, Dr. H.
Pathologisches Institut der Städt. Kliniken Fulda, Pacelliallee 4, 6400 Fulda, West-Germany

Schlaefer, K.
Institut für Dokumentation, Information und Statistik, Deutsches Krebsforschungszentrum, Im Neuenheimer Feld 280, 6900 Heidelberg, West-Germany

Schlüter, Dr. K.
Pathologisches Institut des Städt. Krankenhauses, Robert-Koch-Straße 1, 8300 Landshut, West-Germany

Stegner, Prof. Dr. H.-E.
Univ.-Frauenklinik und -Poliklinik Hamburg-Eppendorf, Martinistraße 52, 2000 Hamburg 20, West-Germany

Trams, Dr. G.
Univ.-Frauenklinik und -Poliklinik Hamburg-Eppendorf, Martinistraße 52, 2000 Hamburg 20, West-Germany

Wittlinger, Prof. Dr. H.
Geburtshilflich-Gynäkologische Abteilung, Krankenhaus Bruchsal, 7520 Bruchsal, West-Germany

Zahradnik, Priv.-Doz. Dr. H.P.
Endokrinologische Abteilung, Klinikum der Albert-Ludwigs-Universität, Univ.-Frauenklinik, Hugstetter Straße 55, 7800 Freiburg, West-Germany

Zieger, Prof. Dr. Gertraud
Pathologisches Institut der Universität des Saarlandes, 6650 Homburg, West-Germany

Zieger, Dr. W.
Pathologisches Institut der Universität des Saarlandes, 6650 Homburg, West-Germany

International Symposium of the Section of Gynecopathology of the German Society of Pathology, Heidelberg, 1979 Opening Remarks

The rapid growth of medical knowledge in the last decade has made it necessary for major fields to split further into subspecialties. Such concentration of study, however, has enabled the specialties of quite different fields to work together more profitably. For example, in the field of pathology functional morphology has been able to develop. The importance of the interplay between morphology and function is nowhere more evident than in the field of gynecology where both are paramount for accurate diagnosis and optimal patient care. Abnormalities in hormone production determine which structural changes occur in the target organs, how the disease manifests itself, and predestine what medical care should be.

Based on the knowledge of these facts we held our first seminar of "gynecological morphology" exactly ten years ago in Mannheim, and that start gave birth to our idea to establish a working group with the same name, where gynecologists and pathologists could meet to discuss common problems. Three years later, after a second seminar in Mannheim in 1971, it was possible to officially establish the working-group for gynecopathology.

Since that time we have regularly held scientific sessions subsidiary to the annual meetings of both parent societies – the German Society of Gynecology and the German Society of Pathology. Since 1975 our working-group has become incorporated as a section into the German Society of Pathology.

It soon became apparent we needed to improve the exchange of information between the disciplines to establish a better bridge between clinical findings and pathological diagnoses. As the use of sex hormones became more widespread, so grew the need for information about structural and functional effects, especially to prevent harmful side effects. The pathologist, however, could not improve his knowledge of hormonal changes without precise clinical information about the patients' use of hormones. In turn, without an exact functional diagnosis from the pathologist, the clinician had difficulty providing optimal therapy.

The purpose of this symposium is to help us exchange information internationally and to foster understanding for each others' problems. To partake in this interchange of ideas we have invited renowned authorities from inland and abroad – from Canada, the United States, Israel, Sweden, France and Spain. Among the participants of the symposium we cordially welcome the guests from Austria, Greece, Italy, Jugoslavia, Portugal, Spain, Switzerland and Hungary.

On behalf of the section of gyneco-pathology and also in commission for Professor Grundmann, president of the German Society of Pathology, I should like to greet you all to Heidelberg most heartily. As honored guests at this time I take special pleasure in welcoming the Rector of the University of Heidelberg, Prof. Niederländer, the Director of the University Institute of Pathology, Prof. Doerr, the Director of the Women's Hospital in Mannheim, Prof. Stoll, and Dr. Prinz from the National Ministry of Research.

In my function as city councillor I am especially delighted and honored to extend to you the personal greetings of our Lord Mayor, Mr. Reinhold Zundel, and to welcome you on behalf of the city of Heidelberg. The official reception by the city will take place this evening in the City Hall. May I express the hope and wish at this time, that besides earnest work you may find time to indulge in our extracurricular activities, in order to see and get to know Heidelberg and its surroundings.

Heidelberg, April 20th, 1979 Gisela Dallenbach-Hellweg

Welcoming Address

Ladies and Gentlemen, my dear Colleagues:

Before the commencement of your conference, please allow a morphologically interested clinician an appeal to the morphological conscience of the pathologist and clinician!

While medicine is striving after the computer that will make diagnoses and even determine therapy for us, the morphologist responsibly arrives at his decisions by casting a glance into the microscope. He does not – and cannot – make his decisions and diagnoses hiding behind a clever piece of apparatus.

True, morphologists can pass a cell through a cannula and determine its properties. But the time is still a long way off when they will succeed in ordering and classifying 10^{14} cells in this manner in the whole human being.

Show now – especially to our rising generation of young medical minds – that commitment, a good training under a good academic teacher, and experience in the medical art always permit of personal decisions for which one bears total responsibility.

Then you will have shown the basis and quintessence of all medical practice.

P. Stoll

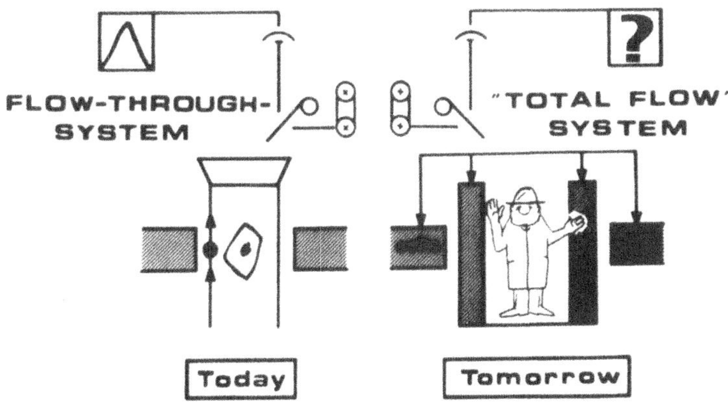

I. Estrogens

The Mechanism of Effect of Estrogens

H.P. Zahradnik and Johanna Beyer

Abstract

In 1960 Jensen and Jacobson provided evidence for a specific binding of estrogens in target organs. This specific binding is the basic requirement for the biological action of estrogens. These actions include stimulation of the synthesis of proteins, for instance enzymes and other steroid hormone receptors.

In addition the data of the present paper indicate that the biological action of estrogens is also involved in the synthesis of prostaglandins being essential for the contractility of the myometrium. The intrauterine application of estradiol valerianate in low doses results in rapid increase of the PGE_2-synthesis. High doses of estrogens, however, are followed by a decreased synthesis of PGE_2.

The biological significance of these results will be discussed in detail.

In 1960 Jensen and Jacobson [1] described a method by which it could be shown that estradiol was bound specifically by the uterus and vagina, while in a number of other organs the radioactivity rapidly disappeared. In other words, nonestrogen target organs take up the radioactive tracer very rapidly and pass it off with equal rapidity. Thus there is no barrier to prevent estradiol from entering the cell. If we look at the subcellular uterus fraction after pretreatment with [^{3}H]estradiol, we find about 60% of the radioactivity in the nuclear fraction and about 30% in the cytosol. Estrogen binding is stereospecific, and [^{3}H]estradiol can be released from the nuclear fraction by proteinases. Analysis by sucrose gradient centrifugation gives us a 9.5 S [^{3}H]estradiol-protein complex in the cytoplasm, and we can extract a smaller 5 S protein complex from the nuclear fraction. This protein is found only in the specific target organs.

If one incubates isolated nuclear fractions from uterine cells with radioactively marked estradiol, one finds that the estradiol is not specifically bound. If one adds uterine cytosol and [^{3}H]estradiol to this incubation medium of isolated uterine nuclei, one can detect the presence of 5 S [^{3}H]estradiol-protein complex. With time we also find an exchange of radioactivity between the cytosol and the nuclear fraction.

Based on the findings of Jensen [2] and Gorski [3] made at about the same time, we can assume that the effect of estrogens on the target organ cells is a two-step process. After the hormone has penetrated the cell membrane, perhaps by simple diffusion, it is bound by the receptors in the cytoplasm and causes their characteristics to change by a process known as acidophilic activation. This activation in turn enables the hormone-receptor complex to react with a number of polyanions in the nucleus, which then, together with other factors, cause the hormone effect.

Natural and synthetic steroids differ only in terms of their biologic potency. The affinity of various estrogens to their corresponding receptors correlates with their ability to stimulate uterine growth, for example. Since a number of protein phosphorilization and dephosphorilization reactions are activated by cyclic nucleotides, the question arises as to what role cAMP or cGMP play in determining steroid effects.

Goldberg [4] administered estradiol to rats and observed lasting changes in the concentration of cyclic nucleotides – an increase in cGMP and a decrease in cAMP. The localization of cGMP, together with the growth-promoting function of cGMP and estrogen, make this observation very interesting, especially since we know that many steroid effects can be mimicked by the cyclic nucleotides.

In general, however, we still do not know what significance the nucleotides have for steroid effects. They surely do not play a role in transferring the cytoplasmic receptor to the nucleus, but they may well be involved in the ensuing estrogen effects.

The two-step model of steroid translocation to the cell nucleus is a satisfying explanation of the mechanism of steroid-controlled enzyme induction. Specific alterations in the cell membrane during the passage of the steroid through the plasma membrane may possibly cause changes in permeability, which are determined by mechanisms that may be unrelated to "de novo" protein synthesis stimulation. The estrogen-related changes in the area of prostaglandins described below, may well fall into this category of estrogen effects.

A series of volunteers ($n = 4$) who presented for treatment of a "missed abortion" gave us blood samples of 20 ml each before their vaginal examination and at certain times after intrauterine instillations of 20, 40, or 60 mg estradiol valerianate. Another group of volunteer patients ($n = 6$) with the same diagnosis served as controls, and were given NaCl solution in physiologic concentrations.

Blood samples were taken before treatment; and 30 s, 2, 10, 30, and 60 min, and 2, 4, and 8 h following intrauterine administration of estradiol valerianate or NaCl solution. Serum concentrations of PGE_2, PGF_{2a}, (DHK-PGF_{2a}) 13,14-dihydro-15-keto-PGF_{2a} were determined radioimmunologically. After intrauterine instillation of 20 mg of estradiol valerianate the concentrations of DHK-PGF_{2a}, PGF_{2a}, and PGE_2 were higher than initial levels and also higher than control levels. The PG concentrations measured by variance analysis following intrauterine NaCl instillation did not change significantly over the entire 8-h period of observation.

An estradiol valerianate dosage of 40 mg did not result in any measurable change in DHK-PGF_{2a} concentration as compared to the controls. The values following treatment were the same as before treatment. Though PGE_2 was significantly higher than the NaCl controls after the administration of 40 mg of estradiol valerianate, its level was clearly lower than after 20 mg. PGF_{2a} levels were significantly higher than the control values following NaCl administration.

After intrauterine instillation of 60 mg estradiol valerianate, the DHK-PGF_{2a} concentrations in the peripheral serum were significantly lower than the corresponding NaCl controls. The PGE_2 values after 60 mg dosage were an average of 60% lower than the control values. The values for PGF_{2a} did not differ significantly from those of the controls on NaCl solution.

Figure 1 may serve to clarify what was said above. Zero on the abscissa corresponds to the mean value of the various DHK-PGF_{2a} concentrations per ml of serum,

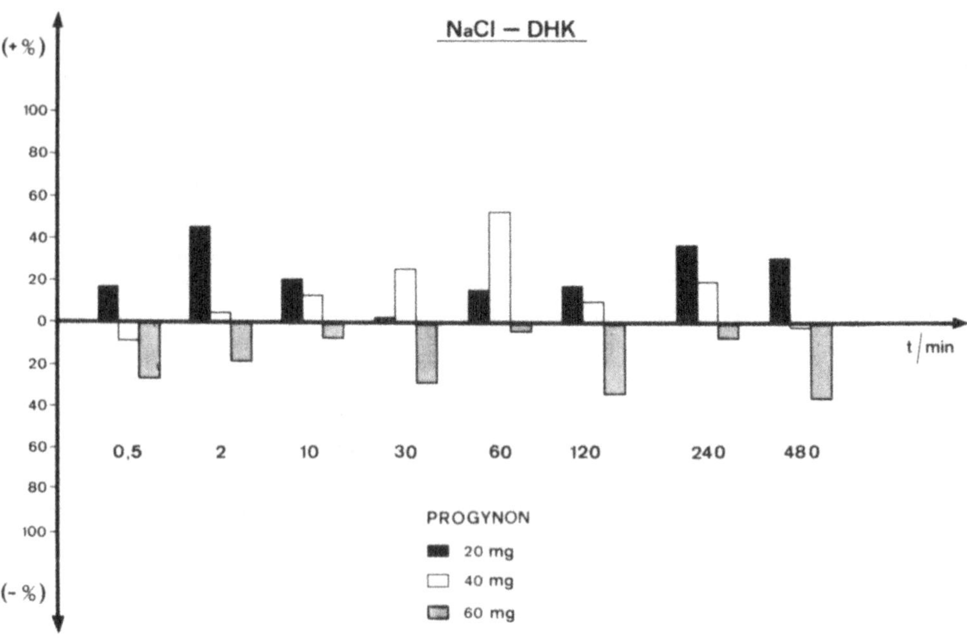

Fig. 1. Changes in 13, 14-dihydro-15-keto-PGF_{2a} (DHK-PGF_{2a}) serum concentration with respect to estrogen dosage / time; the values are given in percent. Zero on the abscissa corresponds to the mean value of the various DHK-PGF_{2a} concentration per ml of serum, following administration of physiological saline

following administration of physiological saline. The columns represent changes in DHK-PGF_{2a} serum concentration with respect to estrogen dosage/time; the values are given in percent. Black columns represent intrauterine dosages of 20 mg, white columns 40 mg, and shaded columns 60 mg of estradiol valerianate. These DHK-PGF_{2a} values show clearly that 20 mg of estradiol valerianate lead to a considerable increase in the stable PGF decomposition product that is measurable in peripheral serum. There is somewhat less of an increase following a 40 mg dosage. However, in all cases in which we administered 60 mg of the drug, we found a significant decrease in the DHK-PGF_{2a} levels in peripheral serum as compared to the controls, whereas the PGF_{2a} levels remained constant.

The nomenclature in figure 2 is the same as in the preceding one, except here we have entered the percentages of PGE_2 serum concentration with respect to estrogen dosage and time. There is a very impressive increase in PGE_2 concentration following intrauterine instillation of 20 mg estradiol valerianate. Expressed in percentages, the values are well over 100% higher than those found in the saline controls. A dosage of 40 mg of the drug also results in a higher PGE_2 serum concentration, though nothing near the effect obtained with the smaller dosage. A 60 mg dosage of estradiol valerianate given intrauterinely results in a significant lowering of PGE_2 concentrations over the entire 8-h observation period, in general by more than 50% as against the corresponding controls on saline solution.

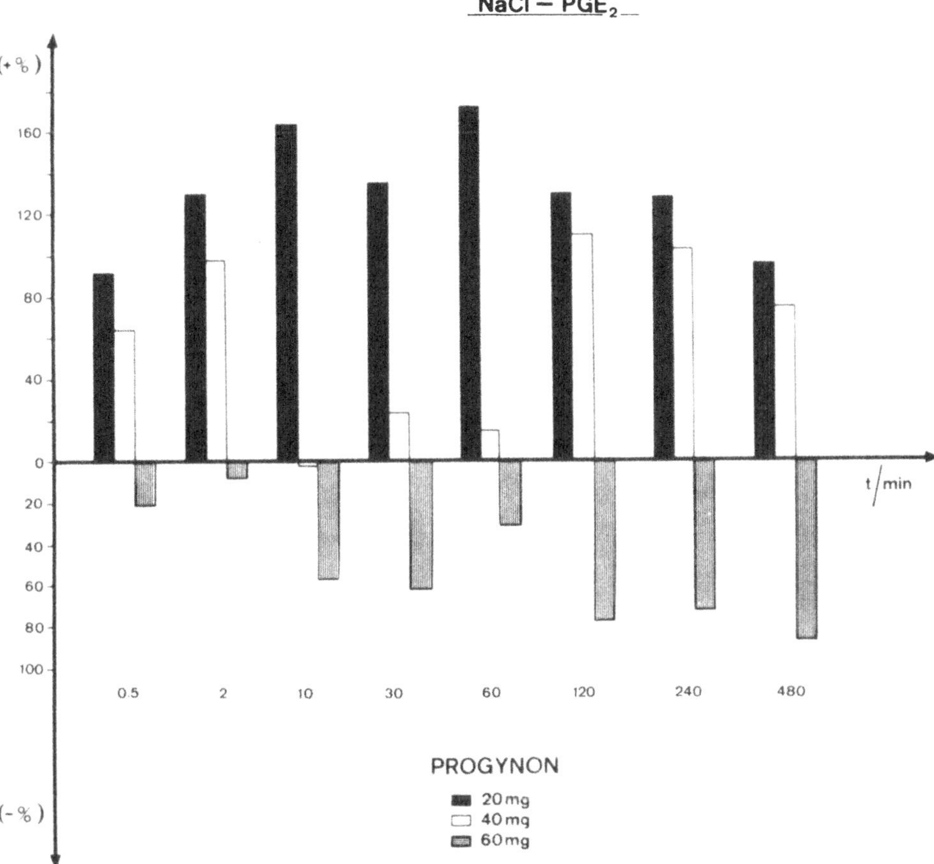

Fig. 2. Changes in prostaglandin E_2 (PGE_2) serum concentration with respect to estrogen dosage / time; the values are given in percent. Zero on the abscissa corresponds to the mean value of the various PGE_2 concentrations per ml of serum, following administration of physiological saline

In their work on the rat uterus, Castracane and co-workers [5] have shown conclusively that steroids do not influence prostaglandin synthesis and decomposition via the usual receptor-DNS-protein synthesis chain, but directly at the cell membrane. Csapo [6] is in agreement with these investigators in pointing out that the progesterone-estrogen imbalance in favor of estradiol is necessary to achieve prostaglandin synthesis. Also, Harney [7] and Cane [8] have pointed out that there is a cooperative effect between PG and estrogens. In ovariectomized rats, they were able to induce uterus contractions with prostaglandins only if the rats had been pretreated with estrogens. These findings are very much in agreement with the results discussed above in the low dosage range. Thus it would seem justified to say that estrogens do indeed influence PGF metabolism, a statement that is supported by the work of several other authors.

However, the increase in PGF and PGE levels does not seem to be dose-dependent, because at higher dosages we find metabolic inhibition. Following the administration of 60 mg estradiol valerianate the DHK-PGF_{2a} and particularly the PGE_2 levels in peripheral serum are significantly lower than the levels found following intrauterine instillation of saline solution. The PGF_{2a} levels by contrast remain in about the same range. What we see, in other words, is a relative predominance of PGF_{2a}.

In sum, the following steroid and prostaglandin interactions may be involved: Up to a certain estrogen concentration in the blood – possibly within the range of physiologic balance – there is an increase in PGF_{2a} synthesis and corresponding decomposition, as indicated by the DHK-PGF_{2a} levels we have found. PGE_2, possibly under the influence of progesterone, can be thought of as protecting the myometrium from contractions that might result from these higher PGF_{2a} concentrations, and is itself present in higher amounts. When the intrauterine dosage of estrogens is further increased – to levels that may be unphysiologic but are biologically meaningful when contractions begin – we produce an absolute imbalance between estrogens and progesterone. We see a definite inhibition of PGF_{2a} decomposition and at the same time an extreme decrease in PGE_2 levels in serum. Possibly this mechanism is the key to the contractive stimulus, which occurs when PGF_{2a} predominates over PGE_2 as a result of the relative predominance of estrogens over progesterone in the myometrium.

References

1. Jensen EV, Jacobson HI (1960) In: Pincus G, Vollmer EP (eds) Biological activities of steroids in relation to cancer. Academic Press, N.Y., p 161
2. Jensen EV, Hurst DJ, De Sombre ER, Jungblut PW (1967) Science 158: 385
3. Shymala G, Gorski J (1967) J Cell Biol 35: 125 A
4. Goldberg ND, Haddox MK, Nicol SE, Glass DB, Sauford CH, Kuehl FA Jr, Estensen R (1975) Adv Cyc Nuc Res 5: 210
5. Castracane VD, Jordan VC (1975) Biol Reprod 13: 587
6. Csapo AJ (1977) Prostaglandins 13: 965
7. Harney PJ, Sueddon JM, Williams KJ (1974) J Endocr 60: 343
8. Cane EM, Villee CA (1975) Prostaglandins 9: 281

The Effect of in Utero Exposure to Diethylstilbestrol

R.J. Kurman and H.J. Norris

The alterations in the lower genital tract associated with diethylstilbestrol (DES) exposure include cervical and vaginal ridges, benign glands in the vagina (adenosis), and exocervix (ectropion) and clear cell carcinoma. (Herbst et al. 1971; Scully et al. 1978; Herbst et al. 1974; Barber and Sommers 1974; Fetherston et al. 1972; Greenwald et al. 1971; Hill 1973; Kanton et al. 1972; Noller et al. 1972; Nordquist et al. 1976; Roth and Hornback 1974; Silverberg and De Giorgi 1972; Tsukada et al. 1972; Herbst et al. 1972; Pomerance 1973; Sandberg 1976; Stafl et al. 1974) These findings have widespread implications since it has been estimated that several hundred thousand young women were exposed to DES in utero in the United States between 1945 and 1970 (Heinonen 1973). Between 1% and 2% of pregnant women received DES during this period. This form of therapy was used mainly in the United States, but it was also used in a limited extent in parts of South America, Africa, and Australia.

Epidemiologic Features

Review of over 300 patients with clear cell carcinoma from the files of the Registry for Research on Hormonal Transplacental Carcinogenesis (Scully et al. 1978; Herbst et al. 1974) indicates that almost two-thirds have a history of maternal DES or chemically related nonsteroidal estrogen such as dienestrol or hexestrol. Patients without a history of in utero DES exposure often represent women whose medical records were incomplete (Scully et al. 1978; Herbst et al. 1974). In some women without a history of exposure it is doubtful the tumor arose spontaneously since clear cell carcinomas and adenosis seldom had been described as arising from the vagina in patients this young prior to the DES era (Hameed 1968; Studdiford 1967). The dosages of hormone administered during pregnancy varied from 1.5 mg to 150 mg daily. In all instances, therapy was initiated in the first half of pregnancy corresponding to the stage of embryogenesis during which the female genital tract develops. Only one clear cell carcinoma has arisen when the drug was started after 18 weeks of gestation.

There are eight cases in the Registry files resulting from exposure to non-DES hormones including steroidal estrogens and progesterone, but these are sporadic cases, a causal relationship between the neoplasm and the steroid hormones having not been demonstrated (Herbst and Cole 1978).

Patients with carcinoma range in age from 7 to 29 years; the mean latency period from the time of exposure to the diagnosis of carcinoma is 19 years. Most of the tumors appear during the period from age 14 to 23 years and less than one woman out of a thousand exposed will develop the tumor in adolescence (Scully et al. 1978).

Benign Lesions Associated with DES Exposure

Almost all women exposed to DES will have one or more of the benign abnormalities such as adenosis, cervical ectropion, and vaginal and cervical deformities. The terms cervical ridges, pseudopolyp (Herbst et al. 1972), strawberry cervix (Pomerance 1973), cervical cock's comblike lesion (Pomerance 1973; Stafl et al. 1974), erythroplakia (Pomerance 1973), vaginal pericervical collar (Herbst et al. 1972), vaginal hood (Pomerance 1973; Stafl et al. 1974), and vaginal band (Barber and Sommers 1974) have been used to describe these deformities. Adenosis is the presence of benign glandular epithelium in the vagina, either in the form of glands lying beneath the vaginal mucosa or, as in its most exaggerated form, it replaces the surface epithelium (Herbst et al. 1972). Although adenosis has been reported in patients in the absence of antenatal DES exposure (Kurman and Scully 1974; Sandberg et al. 1965), clinically detectable adenosis is almost always a result of in utero exposure to DES. Ninety percent of adenosis is located in the anterior upper vagina in fact, confined to this area in almost 90% of cases (Ng et al. 1977) in continuity with cervical ectropion. The natural history of both adenosis and the ectropion is similar in that there is incomplete replacement of the squamous epithelium by glandular epithelium. Both lesions heal by conversion of the glands by immature metaplastic squamous epithelium. This differentiates into mature squamous epithelium indistinguishable from the mature squamous epithelium elsewhere lining the vagina and exocervix. In observations on 75 women, the adenosis resolved in 75% over a period of three years (Antonioli et al. 1979). The zone of active squamous metaplasia intermingled with columnar glandular epithelium (adenosis and

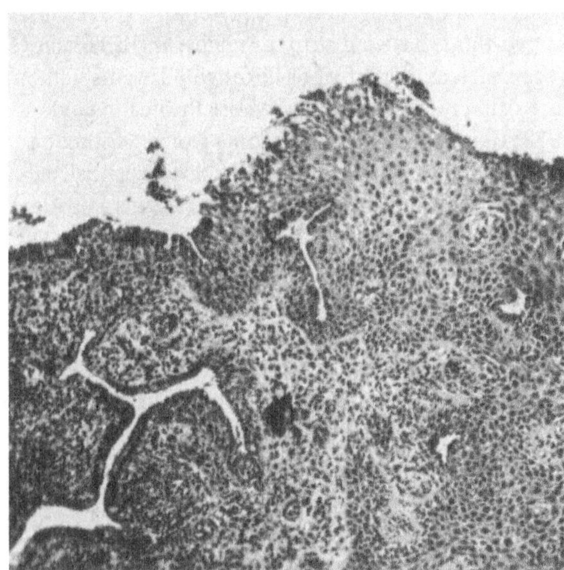

Fig. 1. Vaginal adenosis characterized by glands lying immediately beneath surface squamous epithelium. Note the "transformation" zone present. H and E stain. x 89

ectropion) is analogous to the transformation zone of the cervix (Fig. 1). One of the basic abnormalities of DES exposure is an enlargement of the transformation zone from the cervix into the vagina as well. Since dysplasia and the squamous carcinoma of the cervix arise in the transformation zone, the presence of a greatly enlarged transformation zone within the cervix and upper vagina in DES exposed women prompts speculation that there will be an increase in the incidence of squamous carcinoma in these individuals. Although long term follow-up is still necessary, a study in which 1400 DES-exposed women were examined revealed that the prevalence of vaginal and cervical dysplasia was 2.1% over a 5 year follow-up period (Robboy et al. 1978). The dysplastic changes were almost always mild, whereas severe dysplasia and carcinoma in situ were never encountered (Robboy et al. 1978). Studies utilizing nuclear DNA microspectrophotometry on biopsies referred with a diagnosis of severe dysplasia or carcinoma in situ have shown that an aneuploid distribution of nuclear DNA, consistent with intraepithelial neoplasia, is found in only a small percentage of these cases. The authors conclude that there is an appreciable rate of misinterpretation of squamous lesions occurring in female DES-exposed patients (Richart et al. 1978).

Another benign lesion arising in vaginal adenosis has been termed microglandular hyperplasia and occurs in patients using birth control pills. This is analogous to microglandular hyperplasia of the cervix (pill polyp) occasionally observed in women on birth control pills not exposed to DES. In the vagina, the lesion has been confused both clinically and pathologically with adenocarcinoma (Robboy and Welch 1977; Kurman and Norris 1975). The histologic features are distinctive (Fig. 2) and the lesion is benign requiring no further treatment after excision.

Structural disorders such as cervical and vaginal ridges and septa occur in 22%-58% of the exposed patients (Herbst et al. 1972; Sandberg 1976; Herbst et al. 1975). These generally have little clinical consequence except that their recognition should alert the gynecologist to the possibility of DES exposure. Sometimes a septum must

Fig. 2. Microglandular hyperplasia (right) arising in adenosis. The biopsy is from an 18-years-old woman who had been taking a "high dose" oral contraceptive. H and E stain. x 46

be excised to permit adequate visualization of the cervix and vagina at the time of examination.

Lesions involving the upper genital tract are limited to the uterine cavity. Discovered by hysterosalpingograms, these abnormalities include a T-shaped uterine cavity, uterine synechiae, and hypoplasia in 60% of the patients (Kaufman et al. 1977).

Males have not been immune to the effects of in utero DES exposure. Recent findings indicate that up to 20% of exposed males have had some abnormality and epididymal cysts, hypoplastic testes, cryotorchidism, and abnormalities of sperm including decreased counts, abnormal forms and decreased motility (Gill et al. 1976).

Clear Cell Adenocarcinoma

To judge from the predominant location of the tumor, clear cell carcinoma of the vagina occurs twice as frequently as those arising in the cervix. The vaginal tumors are usually located in the upper third of the vagina on the anterior wall while the cervical tumors are almost always on the exocervix (Scully et al. 1978; Herbst et al. 1974).

The vaginal and cervical tumors are similar in gross appearance, most being polypoid, nodular or papillary, but some flat (Fig. 3) or ulcerated (Scully et al. 1978; Herbst et al. 1974). Microscopically the tumors display solid, papillary, tubular, and cystic patterns and are comprised of clear cells containing abundant glycogen mixed with hobnail cells characterized by scant cytoplasm, a large nucleus protruding into a lumen and flattened cells (Fig. 4). When the tumor consists predominantly of cysts lined by flattened epithelium, its malignant potential is questionable. The identical tumor can occur in the endometrium and ovary, but it has not been associated with DES exposure in these sites and the neoplasms occur in much older women. The ultra-

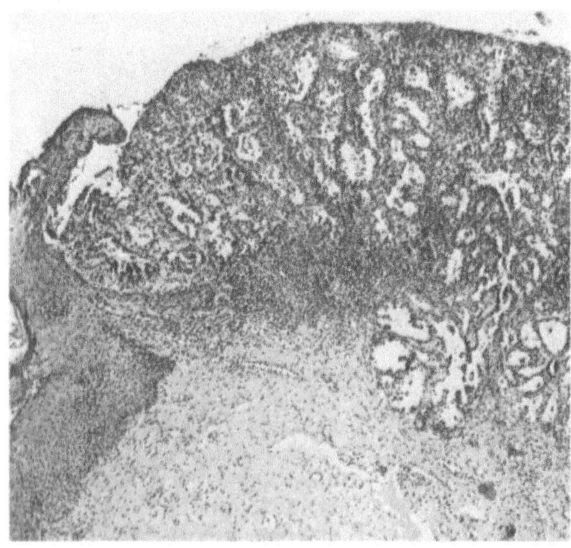

Fig. 3. Low power magnification of a clear cell carcinoma from the lateral vaginal fornix of a 20-year-old woman who had been exposed in utero to DES. A flat lesion, area this neoplasm invaded only 2 mm from the surface. H and E stain. x 82

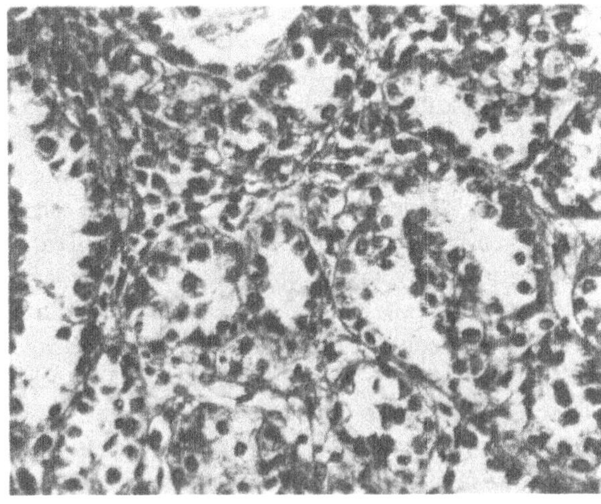

Fig. 4. Higher magnification of the clear cell carcinomas in Fig. 3 showing a mixture of hobnail and clear cells. H and E stain. x 167

structural features of clear cell adenocarcinoma of the cervix and vagina are similar to that reported for clear cell carcinoma of the endometrium and ovary. There are no distinguishing features to the neoplasms associated with DES, and no ultrastructural features are specifically correlated with the depth of invasion or metastasis.

Clear cell carcinoma spreads by local extension, and metastasizes via blood and lymphatic vessels. Approximately 17% of patients with stage I tumors of the cervix and vagina have pelvic lymph node metastasis when first seen (Herbst et al. 1974). Clear cell carcinoma has a greater propensity for distant metastasis than squamous carcinoma originating in the same sites as nearly 35% of recurrent clear cell carcinomas involve the lungs and supraclavicular lymph nodes as compared with only 5%–10% of recurrent squamous carcinoma (Robboy et al. 1974). Factors related to survival include the stage, the size and depth of invasion of the tumor, and, to a lesser extent, its histologic pattern, mitotic activity, and grade. It should be emphasized, however, that although size is an important prognostic feature, pelvic lymph node metastasis has been reported in a tumor with a depth of invasion of 1.5 mm (Herbst et al. 1974).

Pathogenesis of DES Induced Adenosis and Cervical Ridges

The occurrence of adenosis and gross structural abnormalities of the lower genital tract after DES in utero exposure indicates that DES acts as a teratogenic agent altering the normal development of the vagina.

The vagina develops from a solid collection of cells referred to as the vaginal plate. Located at a point where the lower fused portion of the mullerian ducts penetrates the urogenital sinus, the epithelium of the vaginal plate extends cephalad replacing that of that of the fused mullerian ducts and subsequently canalizes; the entire process is completed by the fifth month of gestation (Crobsy and Hill 1962; Forsberg 1973). DES interferes with the replacement of the glandular mullerian epithelium by the squamous epithelium derived from the vaginal plate resulting in the sporadic presence of aberrant glandular epithelium in the vagina and cervix (Forsberg 1960). Gland-like downgrowths into the stroma of the vagina of neonatal mice have also been produced by the administration of estradiol-17B and DES (Forsberg 1972). Since it has been shown in these animals that the underlying stroma can influence the epithelium in the vagina (Cuhna 1976), it is conceivable that DES may exert part of its primary effect on this stroma accounting for the various gross structural abnormalities observed. At times the drug has been administered prior to the recanalization of the vaginal plate so DES can alter the expression of the potential of future cells by acting on antecedants. The precise role of DES as a carcinogen is not as clear as its capacity as a teratogenic agent. Clear cell carcinoma, being an adenocarcinoma, should arise from glandular epithelium. Since the vagina is normally devoid of glands, and adenosis is present closely admixed with the tumor in 95% of cases, it is generally accepted that clear cell carcinoma arises from adenosis. Although adenosis is very common in the exposed population, the development of carcinoma is quite rare. Furthermore, the tumors appear to develop over a comparatively restricted interval, the incidence rising sharply from age 14 to a peak of 19 years and then declining through age 22. The rise in incidence at age 14, corresponding with the onset of puberty, suggests that hormonal stimulation plays a role. DES may act primarily as a teratogenic agent producing adenosis, a necessary precursor for the development of clear cell carcinoma. Stimuli which do not effect the normal squamous epithelium of the vagina may be capable of inducing a neoplastic change in glandular tissue normally not present in this location.

Clinical Evaluation of the DES Exposed Individual

Women exposed to antenatal DES must be examined starting either at the time of menarche or by the age of 14 years if menarche has not occurred. Examination should

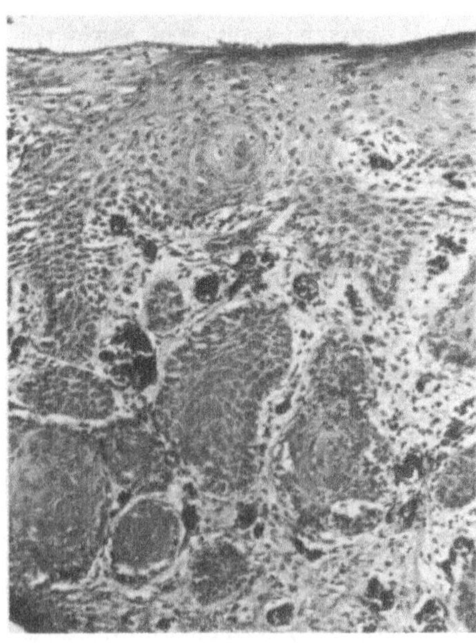

Fig.5. Metaplastic squamous epithelium replacing mucinous epithelium of vaginal adenosis in the anterior vagina of a 20-year-old woman. Eventually the deeply placed squamous nests may disappear. H and E stain. x 96

Table 1. Therapy and survival of 297 patients with clear cell carcinoma[a].

Location and stage	Number of patients	Number with recurrence	Number dead	% Living free of disease
Vagina				
Stage I	129	16	10	80%
Stage II	36	2	6	17%
Stage III and IV	11	2	9	–
Sub total	176 (100%)	20 (11%)	25 (14%)	
Cervix				
Stage I	46	1	4	90%
Stage IIA	43	6	7	70%
Stage IIB	20	0	9	55%
Stage III and IV	12	2	8	17%
Sub total	121 (100%)	9 (7%)	28 (23%)	
Total:	297 (100%)	29 (9%)	53 (18%)	

[a] Data from the Registry for Research on Hormonal Transplacental Carcinogenesis (Herbst and Cole 1978)

include inspection of the entire circumference and length of the vagina and cervix, supplemented by iodine staining and colposcopy. The latter may be confusing, however, since the extensive immature squamous metaplasia which occurs in these patients makes interpretation difficult. In view of the imprecise correlation between the colposcopic and the histologic findings in evaluating adenosis, it is felt by some colposcopist (Townsend 1978) that this technique is best utilized to assess the extent of the epithelial change rather than as a means for directing biopsies. Small tumors have been detected by palpating for areas of induration.

Cytologic examination is an important screening technique. Positive or suspicious smears have been reported in 76% of patients with tumors (Taft et al. 1974). More thorough collection techniques, utilizing four quadrant scrapings of the entire length of the vagina, can detect adenosis in 69% of patients, similar to the frequency of adenosis detected by colposcopy (Ng et al. 1975; Bibbo et al. 1975).

Biopsy should be performed on indurated or cystic, discrete lesions that appear red, granular or ulcerated, and on areas with highly abnormal colposcopic patterns (atypical vessels or heavy mosaic and punctation) (Townsend 1978). Depending on the findings at the initial examination, follow-up should be at 6 to 12 month intervals.

Treatment and Survival

At least a fourth of young women with clear cell carcinoma of the vagina and cervix will develop recurrences (Herbst et al. 1974) but this figure must still be regarded as tentative in view of the relatively short follow-up period (Table 1). Two patients have developed recurrence after a seven year disease free interval, emphasizing the need for prolonged follow-up (Herbst and Cole 1978).

Both radical surgery and radiation have been used effectively for patients with stage I vaginal tumors and stage I and IIa cervical neoplasms (Herbst et al. 1974). Local resection has been advocated for small, superficial carcinomas, but 17% of patients with stage I tumors of the cervix and vagina have pelvic lymph node metastasis and metastasis has occurred with tumors which have invaded only 1.5 millimeters (Herbst et al. 1974). Thus, at the present time, wide local excision cannot be recommended.

Local pelvic recurrence has been successfully controlled with both surgery and radiation. The results with chemotherapy have been disappointing. Multiple chemotherapeutic agents are presently being evaluated for control of disseminated disease. These include cyclophosphamide (Cytoxan), vincristine, 5-fluorouracil, methotrexate, and prednisone (Herbst and Cole 1978).

References

1. Antonioli DA, Burke L, Rosen S (1979) Evaluation of diethylstilbestrol associated lower genital tract abnormalities. Lab Invest 40:231
2. Barber HRK, Sommers SC (1974) Vaginal adenosis, dysplasia, and clear-cell adenocarcinoma after diethylstilbestrol treatment in pregnancy. Obstet Gynecol 43: 645–682
3. Bibbo M, Ali I, Al-Naqeeb M, Baccarini I, Climaco L, Gill W, Sonek M, Weid GL (1975) Cytologic findings in female and male offspring of DES-treated mothers. Acta Cytol (Baltimore) 19:568–572
4. Crosby WM, Hill EC (1962) Embryology of the Mullerian duct system. Obstet Gynecol 20:507–515
5. Cuhna GR (1976) Stromal induction and specification of morphogenesis and cytodifferentiation of the epithelia of the Mullerian ducts and urogenital sinus during development of the uterus and vagina in mice. J Exp Zool 196:361–370
6. Fetherston WC, Meyers A, Speckhard ME (1972) Adenocarcinoma of the vagina in young women. Wis Med J 71:87–93
7. Forsberg JG (1960) Effect of sex hormones on the development o the rat vagina. Acta Endocrinol (Kbh) 33:520–531
8. Forsberg JG (1972) Estrogen, vaginal cancer and vaginal development. Am J Obstet Gynecol 113:83–91
9. Forsberg JG (1973) Cervicovaginal epithelium: Its origin and development. Am J Obstet Gynecol 115: 1025–1043
10. Gill WB, Schumacher GFB, Bibbo M (1976) Structural and functional abnormalities in the sex organs of male offspring of mothers treated with diethylstilbestrol (DES). J Reprod Med 16:147–152
11. Greenwald P, Barlow JJ, Nasca PC, Burnett WS (1971) Vaginal cancer after maternal treatment with synthetic estrogens. N Engl J Med 285:390–392
12. Hameed K (1968) Clear-cell "mesonephric" carcinoma of the uterine cervix. Obstet Gynecol 32:564–575
13. Heinonen OP (1973) Diethylstilbestrol in pregnancy. Frequency of exposure and usage patterns. Cancer 31:573–577
14. Herbst AL, Cole P (1978) Epidemiological and clinical aspects of clear-cell adenocarcinoma in young women. In: Herbst AL (ed) Intrauterine exposure to diethylstilbestrol in the human. Proceedings of Symposium on DES, 1977. The American College of Obstetricians and Gynecologists, Chicago, Ill., pp 2–7

15. Herbst AL, Ulfelder H, Poskanzer DC (1971) Adenocarcinoma of the vagina: Association of maternal stilbestrol therapy with tumor appearance in young women. N Engl J Med 284:878–881
16. Herbst AL, Kurman RJ, Scully RE (1972) Vaginal and cervical abnormalities after exposure to stilbestrol in utero. Obstet Gynecol 40:287–298
17. Herbst AL, Robboy SJ, Scully RE, Poskanzer DC (1974) Clear-cell adenocarcinoma of the vagina and cervix in girls: Analysis of 170 registry cases. Am J Obstet Gynecol 119:713–724
18. Herbst AL, Poskanzer DC, Robboy SJ, Friedlander L, Scully RE (1975) Prenatal exposure to stilbestrol. A prospective comparison of exposed female offspring with unexposed controls. N Engl J Med 292:334–339
19. Hill EC (1973) Clear-cell carcinoma of the cervix and vagina in young women. A report of six cases with association of maternal stilbestrol therapy and adenosis of the vagina. Am J Obstet Gynecol 116:470–484
20. Kanton HI, Weinstein SA, Kaye HL (1972) Clear-cell adenocarcinoma in young women. Obstet Gynecol 41:443–446
21. Kaufmann RH, Binder GL, Grav PM, Jr. Adam E (1977) Upper genital tract changes associated with exposure in utero to diethylstilbestrol. Am J Obstet Gynecol 128: 51–56
22. Kurman RJ, Norris HJ (1975) Letter to the editor: Microglandular hyperplasia superimposed on vaginal adenosis. Obstet Gynecol 46:373–374
23. Kurman RJ, Scully RE (1974) The incidence and histogenesis of vaginal adenosis. Hum pathd 5:265–276
24. Ng ABP, Reagen JW, Hawliczek S, Wentz WB (1975) Cellular detection of vaginal adenosis. Obstet Gynecol 46:323–328
25. Ng AB, Reagen JW, Nadji M (1977) Natural history of vaginal adenosis in wcmen exposed to diethylstilbestrol in utero. J Reprod Med 18:1–13
26. Noller KL, Decker DG, Lanier AP, Kurland LT (1972) Clear-cell adenocarcinoma of the cervix after material treatment with synthetic estrogens. Mayo Clin Proc 47:629–630
27. Nordquist SRB, Fidler WJJr, Woodruff JM, Lewis JLJr (1976) Clear-cell adenocarcinoma of the cervix and vagina. A clinicopathologic study of 21 cases with and without a history of maternal ingestion of estrogens. Cancer 37:858–871
28. Pomerance W (1973) Post-stilbestrol secondary syndrome. Obstet Gynecol 42:12–18
29. Richart RM, Fu YS, Reagen JW, Barron BA (1978) The problem of squamous neoplasia in female DES progeny. In: Herbst AL (ed) Intrauterine exposure to diethylstilbestrol in the human. The American College of Obstetricians and Gynecologists, Chicago, Ill., pp 45–52
30. Robboy SJ, Welch WR (1977) Microglandular hyperplasia in vaginal adenosis associated with oral contraceptives and prenatal diethylstilbestrol exposure. Obstet Gynecol 49:430–434
31. Robboy SJ, Herbst AL, Scully RE (1974) Clear-cell adenocarcinoma of the vagina and cervix in young females: Analysis of 37 tumors that persisted or recurred after primary therapy. Cancer 34:606–614
32. Robboy SJ, Keh PC, Nickerson RJ, Helmanis EK, Prat J, Szyfelbein WM, Taft PD, Barnes AP, Twat EZ, Scully RE, Welch WR (1978) Squamous cell dysplasia and carcinoma in situ of the cervix and vagina after prenatal exposure to diethylstilbestrol. Obstet Gynecol 51: 528–535
33. Roth LM, Hornback NB (1974) Clear-cell adenocarcinoma of the cervix in young women. Cancer 34:1761–1768
34. Sandberg EC (1976) Benign cervical and vaginal changes associated with exposure to stilbestrol in utero. Am J Obstet Gynecol 125:777–788
35. Sandberg EC, Danielson RE, Cauwet RW, Bonar BE (1965) Adenosis vaginae. Am J Obstet Gynecol 93:209–222
36. Scully RE, Robboy SJ, Welch WR (1978) Pathology and pathogenesis of diethyl-

stilbestrol-related disorders of the female genital tract. In: Herbst AL (ed) Intrauterine exposure to diethylstilbestrol in the human. Proceedings of Symposium on DES, 1977. The American College of Obstretricians and Gynecologists. Chicago, Ill., pp 8–22
37. Silverberg SG, DeGiorgi LS (1972) Clear-cell carcinoma of the vagina: A clinical, pathologic, and electron microscopic study. Cancer 29:1680–1690
38. Stafl A, Mattingly RF, Foley DV, Fetherston WC (1974) Clinical diagnosis of vaginal adenosis. Obstet Gynecol 43:118–128
39. Studdiford WE (1967) Vaginal lesions of adenomatous origin. Am J Obstet Gynecol 73:641–656
40. Taft PD, Robboy SJ, Herbst AL, Scully RE (1974) Cytology of clear-cell adenocarcinoma of the genital tract in young females: Review of 95 cases from the registry. Acta Cytol (Baltimore) 18:279–290
41. Townsend DE (1978) Techniques of examination and screening of the DES-exposed female. In: Herbst AL (ed) Intrauterine exposure to diethylstilbestrol in the human. Proceedings of the Symposium on DES, 1977. The American College of Obstetricians and Gynecologists, Chicago, Ill., pp 23–29
42. Tsukada Y, Hewett WJ, Barlow JJ, Pickren JW (1972) Clear-cell adenocarcinoma ("Mesonephroma") of the vagina: Three cases associated with maternal synthetic nonsteroid estrogen therapy. Cancer 29:1208–1214

On the Teratogenic Action of Diethylstilbestrol and Other Exogenous Sexual Hormones

B. von Schilling

Abstract

The synthetic estrogen diethylstilbestrol used between 1948 and 1970, mainly in the United States, as an antiabortive agent was found to be the cause for the appearance of vaginal clear-cell adenocarcinoma in around 350 young women born to mothers who were treated with this product during pregnancy because of the threat of abortion. The discovery of this etiologic relationship was made by a group of gynecologists and pathologists in Boston after seeing a clustering of the first eight cases within 3 years (Herbst et al. 1971). The pathogenetic mechanism of this cancer in women prenatally exposed to DES has since been understood by toxicologists as a transplacental cancerogenesis – the only one found as far in the human being.

This theory is discussed in view of the fact that cancer developed only in a very low percentage of such prenatally exposed young women while a large majority presented with a benign teratogenic change: vaginal adenosis. To the pathologist the dysplastic transformation of glands of this atypical vaginal mucosa into adenocarcinoma in the pubertal woman suggests itself as a more likely pathogenesis. Thanks to the continuous investigation this hypothesis has now obtained strong support by casuistic material with evidence of such direct transformation.

Apart from the masculinizing effects found in daughters of women treated with progestins against premature delivery or abortion, DES thus is the only sexual hormonelike synthetic drug with a proven teratogenicity. All other sexual steroids used in contraception, pregnancy tests and so-called protective treatment have not as yet provided clear cut evidence for a teratogenic action.

Introduction and Main Findings

The sudden observation of the very rare clear-cell adenocarcinoma in the vagina of eight girls aged 18 to 22 years within 3 years (1966–1969) prompted Herbst, Ulfelder, and Poskanzer (Herbst and Scully 1970; Herbst et al. 1971) to search for a common environmental cause. To this end the histories of the girls and their mothers were matched with 32 other mother daughter pairs (four per patient); the mothers came from the same geographic area and gave birth at the same hospital in Boston within 5 days before or after the birth of the patients.

Three features stood out for their similarity among the mothers of the affected girls and were significantly absent in the control mothers:

1. The pregnancies of the propositi were complicated by spotting or bleeding.
2. Former pregnancies of the same mothers had ended with fetal loss.

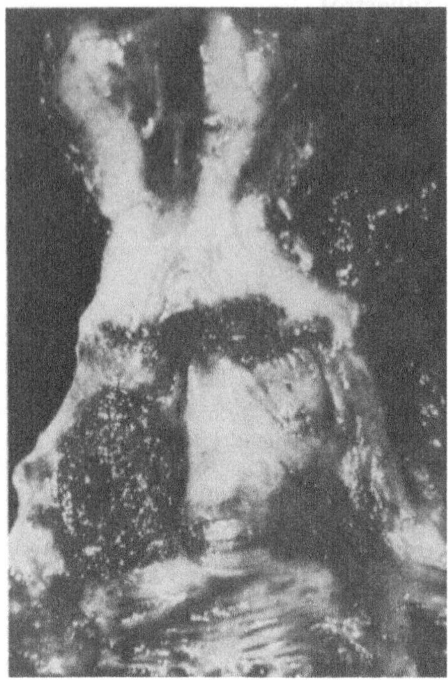

Fig. 1. Operative specimen of radical vagino-hysterectomy with clear-cell adenocarcinoma at right edge and adenosis vaginalis at cervix and left edge of the ventrally opened vagina. (Published by Herbst et al. 1970 Herbst et al. 1972 Herbst et al. 1975 and Ulfelder 1976a; present copy reproduced from lantern slide kindly provided by R.E. Scully)

3. Seven of eight mothers had been treated during the first trimester of their pregnancy with the nonsteroidal synthetic product diethylstilbestrol (DES), the estrogenlike action of which was first described by Dodds in 1938 (cit. by Goldstein 1974) and which was recommended for the protective treatment of high-risk pregnancies in 1948 by Smith (1948).

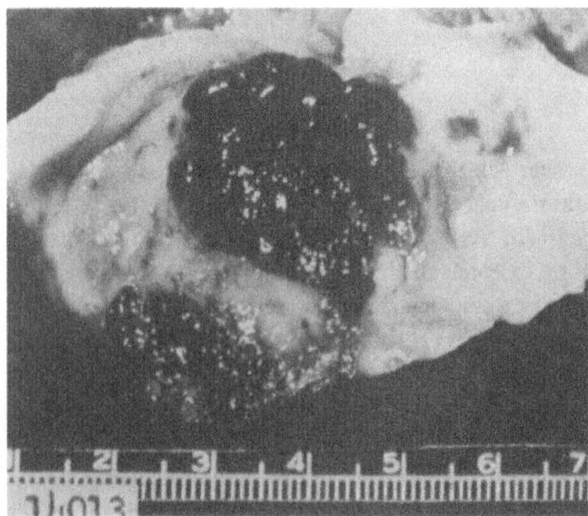

Fig. 2. Aspect of vaginal clear-cell adenocarcinoma. (Scully, unpublished information).

Fig. 3. Low power view of vaginal tumor invasion. (Scully, unpublished information).

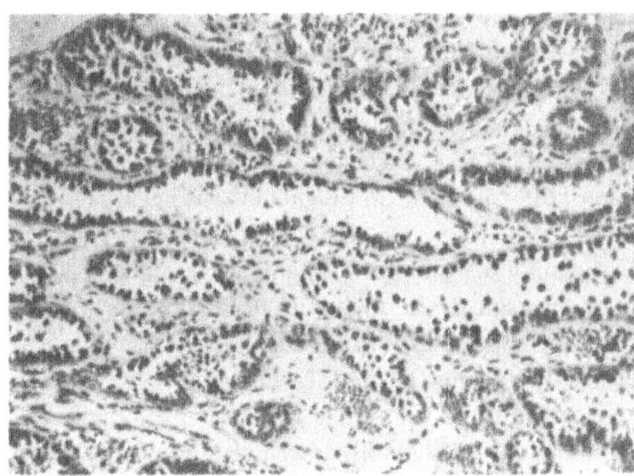

Fig. 4. "Hob-nail" cell lining of adenocarcinomatous tubules. (Scully, unpublished information).

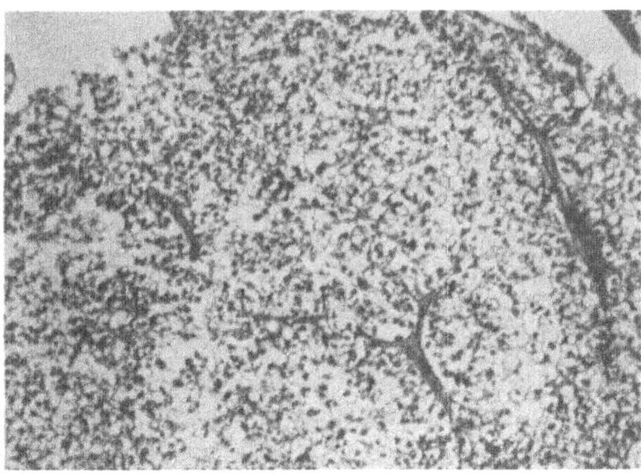

Fig. 5. Clear cells proliferating from fine capillary stalks. (Scully, unpublished information).

The patients presented with fungating tumors originating principally from the anterior wall of the vagina, some of them extending as far down as the urethral orifice. Microscopically they were identified as clear-cell adenocarcinomas with hob-nail like cells lining glandular or pseudoglandular tubules and with glycogen-rich clear-cells proliferating from a scanty stroma (Figs. 1–5).

Through the initiative of Herbst and Scully (Herbst, 1976) a registry of clear-cell adenocarcinoma (mesonephroma) of the genital tract in young female, was set up in 1971 to spread the notice of this iatrogenic accident and to reduce the damage and the risk by the early discovery of more of such tumors in girls exposed to DES in utero.

Other Findings

Already in these first patients a nearly constant association was noted of the neoplastic finding with other nontumorous abnormalities in the genital tract: Transverse fibrous ridges or vaginal hoods located high up in the vagina obstructed the inspection of the cervix region; cervical ectropion was frequent and there was a high incidence of the so-called vaginal adenosis. This latter condition was first described in autopsy material by von Preuschen in 1877. Herbst et al. (1974) observed it together with the vaginal neoplasm in up to 97% of 76 cases published in 1974. Its general incidence increased in an almost linear fashion, the earlier that DES-treatment in pregnancy was begun. (Kurman 1979).

Adenosis is understood as a congenital disturbance in vaginal development during the period of the squamous lining of the vagina. The squamous epithelium pushing upwards from the vaginal plate seems to be kept by DES from substituting the glandular tissue remnants of the fused Müllerian ducts (Herbst 1976; Herbst et al. 1975). Therefore focal patches of columnar cell epithelium line the vagina or form glandular tubules that spread underneath a thin squamous layer. Endometrial or salpingian type epithelium with cilia has also been observed with some frequency in these foci.

Vaginal adenosis and cervical ridges *without* the additional carcinoma proved to be far more frequent in all following reviews of DES-exposed girls. Hart et al. (1976) observed it in 43 of 80 young females in a study realized by the Stilbestrol Registry of the University of Southern California, on girls believed to have had prenatal exposure to DES (Figs. 6–9).

Malignancy

The short distance between the neoplastic mucosal cells and the lymphatic vessels in the underlying lamina propria explains the poor prognosis of this carcinoma, similar to the classic squamous vaginal carcinoma in older women. The natural hesitation to submit young girls with only slight symptoms of vaginal bleeding or discharge to a thorough gynecologic instrumental examination has frequently caused delay in the detection of the tumor and thus increased mortality.

After the review of 170 cases collected by the Registry in 1974, Herbst et al. (1974) found positive lymph nodes in 4 of 12 vaginal and 2 of 7 cervical clear-cell ade-

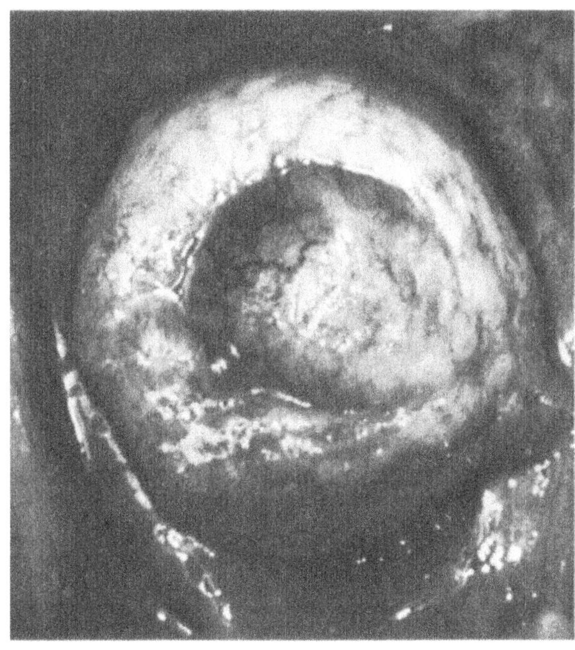

Fig. 6. Cervical ridge or vaginal hood frequently associated with prenatal intra-uterine exposure to DES. (Scully, unpublished information).

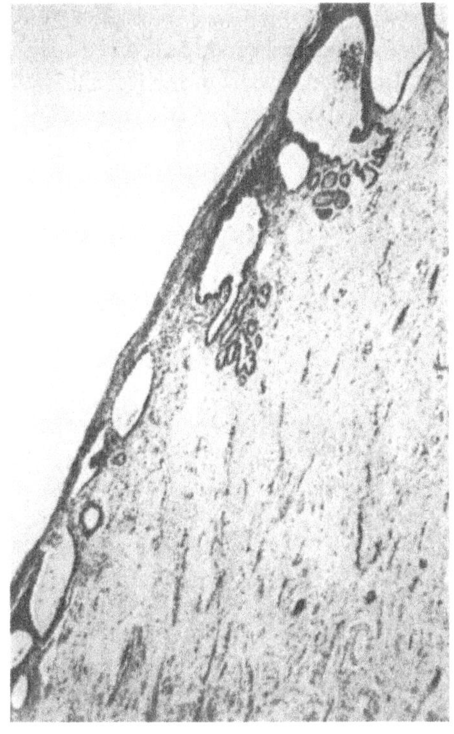

Fig. 7. Low power view of adenosis vaginalis. (Herbst and Scully 1970)

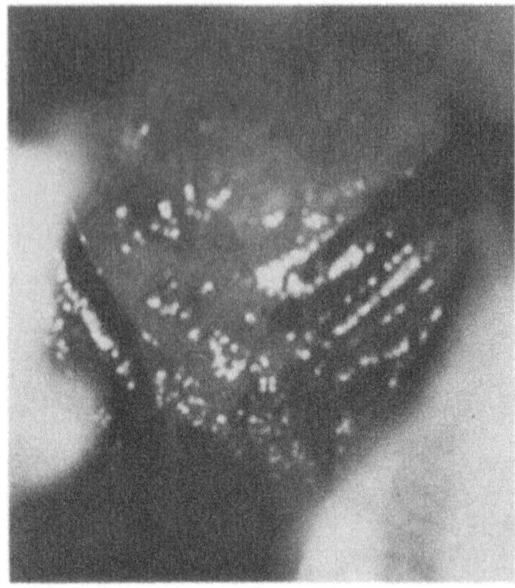

Fig. 8. Jodine negative mucosal areas in vaginal adenosis. (Scully, unpublished information)

nocarcinomas of stage I. 24% of 154 patients, whose tumors could be staged, had died. Prognosis therefore depended mainly upon the development the tumor hat attained upon detection. (Greenwald et al. 1971).

Incidence (Females)

In 1979, Kurman stated that the total number of vaginal and cervical clear-cell adenocarcinomas presently registered was above 300. However, in view of the fact that between 100,000 and 160,000 live born females were exposed to DES in utero between

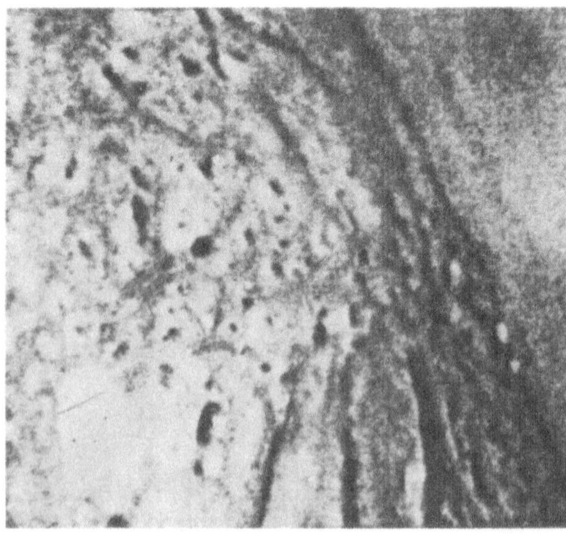

Fig. 9. Vaginal mucosa with vaginal adenosis showing gland openings. White areas to the left diagnosed as squamous dysplasia. (Fetherstone 1975)

1960 and 1970 in the United States, it appeared that probably less than one exposed woman out of one thousand would develop carcinoma in adolescence. In nearly all his papers Herbst emphasizes the exceedingly low incidence of clear-cell adenocarcinoma when compared to the considerable percentages of the other benign abnormalities associated with intrauterine exposure to DES. The malignant neoplasm is thus the rare and extreme expression of a complex teratogenic lesion caused by DES or other structurally alike, nonsteroidal, synthetic estrogens in the human female when administered to the mother before the 21st week of pregnancy.

Incidence outside of the United States:
Monaghan et al. (1978) published the first reported case of vaginal clear-cell carcinoma in the United Kingdom that occurred in a 25 year old woman exposed prenatally to DES between the 5th and 14th week.

Incidence (Males)

In the human *male* born after identical intrauterine DES-exposure Bibbo et al. described epididymal cysts in 10% of 42 exposed males in 1975. In 1976, Gill et al. reviewed 134 males and found 19 epididymal cyst carriers, 9 males with hypotrophic testis, 3 with hypoplastic penis, altogether 36 that compare with 8 such findings among the 119 controls. Blood-hormone assays were within normal range. Severely pathologic changes in spermatozoa analysis were reported in 29% of 28 exposed males. There were no neoplasms. In 1977 Bibbo et al. again report on these same findings. In the same year Cosgrove (1977) made a mail-questionnaire study on 225 DES-exposed men compared with 111 controls. At the beginning of this paper he remarks that the mothers had also received other estrogens during their pregnancy and more progestin than the control mothers. He observed a total of ten penile abnormalities and eight other genitourinary findings plus three undescended testis, one hypospadia, and four meatal stenosis compared to four such lesions among the controls. The possible appéarance of lesions in the Müllerian derived parts of the male genital organs (prostata, utriculus) in the same men in later age, is discussed. The same authors also refer to the findings of McLachlan (1975) of hypoplastic and preneoplastic changes in the male genital organs of mice after prenatal DES-exposure.

Pathogenetic Hypothesis

All the reported findings support a doubtless teratogenic action of DES. This was suspected by Herbst et al. in 1971 and formulated in 1972 (Herbst et al.). The editor of the New England Journal of Medicine highlighted the first article of this discovery and with an acute sense for its importance forecast its tremendous implication for research and public health (Langmuir 1971).

The experimental production of vaginal adenosis was tried by Herbst himself (1976). He states that in his DES treated rats indeed a columnar type epithelium was present in the vagina but also in some of the control rats. So he is hesitant on extrapolation from this result to man. Forsberg (1972) injected newborn mice (SC) with

DES 24 h after their birth and once daily during the next five days. He shows figures of columnar epithelium remnants in the fornices of the vagina. Seven weeks after birth, at the time of sexual maturity, the DES-injected females showed more extensive regions of columnar epithelium in the cervical canal and in the vaginal fornices and a marked tendency of the columnar epithelium to push glandular downgrowths into the adjacent stroma. No detailed information is given on the number and on the findings of his control mice.

Forsberg claims a pronounced mitotic inhibition provoked by DES in the untransformed pseudocolumnar epithelium of the anterior part of the vagina-anlage and of the uterine cervix. It seems understandable that in these regions the pending lining with squamous mucosa also stays behind instead of removing and substituting the glandular tissue of the Müllerian duct.

Dose and Time Dependence

Dose and duration of stilbestrol administration varied highly among the mothers registered by the Boston authors so that no convincing dose dependence could be established for the occurrence of either benign or malignant findings in the female offspring. – There were no findings in the offspring of mothers treated after the 21st week of pregnancy.

Transplacental Carcinogenesis

In 1971, Folkman states that the detection of vaginal adenocarcinomas in correlation with maternal ingestion of DES would be the first demonstration of transplacental carcinogenesis in man:

"On the assumption that only a single fetal cell in the future vaginal tissue undergoes malignant transformation because of stilbestrol, this genetic defect might not be disclosed during the prepubertal years, when cellular renewal is sluggish and only a rare cell is undergoing mitosis. At puberty the vaginal epithelium responds to surges of hormonal stimulation and depletion. Millions of cells are swept into the mitotic cycle but always return obediently to the differentiated state. It is possible that this burst of mitotic activity provides the opportunity for a lurking defective cell to display its malignant nature . . .
. . . The hormonal stimulation of puberty in a sense may be the *promoter*; stilbestrol may be the *initiator*."

Perhaps such a plausible hypothesis explains the readiness with which toxicology ever since has linked the action-profile of DES with the concept of transplacental carcinogenesis (Frohberg 1977, 1978) although for DES this mechanism has not yet been reproduced in experimental animals.

In his treatise on prenatal carcinogenesis Ivankovic (1975) interpretes the findings of Herbst as a confirmation in humans of his and other's experimental demonstration that a transplacental production of malignant tumors with long latencies of manifestation in the offspring is possible. (Malignant tumors of the nervous system in the offspring of nitrosoamine treated dams).

It seems noteworthy, however, that such tumors in rodents only could be produced when the mothers received the compounds during the very last days of their pregnancy. This is in contrast with the sensible period of the human embryo for DES. During the last days of fetal development the nervous tissue of rodents should have reached a high grade of cellular differentiation. In the human embryo during the first three pregnancy months, which was the suitable period for the teratogenic action of DES, the affected structures derived from the Müllerian ducts will still consist mainly of undifferentiated cells, which is a rather distinct anatomical condition, that to our mind restrict the conclusions of analogy.

The Boston group of authors formulate their pathogenetic hypothesis in a more reserved way. In 1976, Ulfelder comments that the primary effect of DES, i.e., vaginal adenosis, cervical ridges, and other abnormalities, seems clearly to be teratogenous and consequent upon stilbestrol exposure:

"Whether the infrequent late adenocarcinoma is stilbestrolinduced, however, can only be speculation at this time. Careful sequential prospective observation of the women with adenosis and also of animal models should help identify the serial steps of the progression in those that ultimately develop neoplasms.

As benign adenosis generally is present when carcinoma is found, tissue changes intermediate between adenosis and adenocarcinoma ought to be assumed to be the probable sequence of events. Such transition, however, has not yet been proved in these cases."

Here for the first time the possibility of a transformation of a vaginal adenosis into adenocarcinoma is discussed and "transition" in such a set-up should mean the insurgence of dysplastic changes, which in other human adenocarcinomas are a common predecessor.

However, up to 1979 there seems to have appeared only one paper in which dysplastic columnar cells in glandular tubules were demonstrated in two cases of vaginal adenosis in DES-exposed girls (Barber and Sommers 1974). The authors comment that the development of vaginal clear-cell adenocarcinoma from vaginal adenosis appears evident. However, with these cases they do not yet claim to have shown a direct connection between the observed dysplasia and the formation of clear-cell adenocarcinoma because preliminary studies on the nuclear DNA in adenosis compared to adenocarcinoma had revealed only diploid or tetraploid types in adenosis and aneuploidy in adenocarcinoma.

After the present paper was submitted for publication (July 1979) Scully informed personnally about Antonioli's et al. recent paper (1979) in which these described true dysplastic changes in glandular tubules of vaginal adenosis in a 21 year old nulliparous woman prenatally exposed to DES. By means of serial cuts severe dysplasia could be proved in parts of the lining of the same tubules in which other areas still conserved a regular columnar lining of the endometrioid type. This dysplastic transformation "may represent tissue in an intermediate phase in the development of clear-cell adenocarcinoma."

Teratogenicity of Other Exogenous Sexual Hormones

After a series of papers on teratogenic effects of exogenous female sexual hormones used for contraception, pregnancy tests, or prevention of threatened abortion had been published from 1973 onward (Nora and Nora 1973; Janerich et al. 1974; Janerich 1975; Harlap et al. 1975) considerable disquietness was caused among the medical profession eliciting a noteworthy editorial in The Lancet (1974) which tried to answer the fundamental question whether sex hormones were teratogenic. This editorial discusses the four apparently unrelated anomalies (limb reduction, Janerich; congenital defects of the great vessels, Nora; and malformations of *v*ertebral, *a*nal, *c*ardiac, *t*racheal, *e*sophageal, *r*enal, *l*imb structures, resumed under the contraction Vacterl and the DiGeorge's syndrome, Nora and Nora) linked with hormone ingestion in early pregnancy and draws attention to the fact that some of these studies were of conflicting evidence because the drug inquiries were generally retrospective and sometimes effected by telephone:

"In retrospective studies, the mothers of the malformed are more ready than controls to give a positive history of almost anything." In second place, this editorial dwells on the fact that the described malformations were remarkably unspecific, comparable only to antimitotic drug effects and quite unlike to thalidomide. "With the exception of the undisputed masculinization of female fetuses by some progestational drugs (Wilkins et al. 1958, Wilkins 1960, Revesz 1960) the evidence linking hormone ingestion in early pregnancy with malformation is so far extremely tenuous."

Since this editorial a huge prospective epidemiologic study on 50.282 probands has been realized by the United States collaborative perinatal project published by Heinonen et al. (1977) in a book of more than 500 pages.

According to a detailed and critical study of this report by Nocke (1978) a statistically significant teratogenous action by hormones, hormone-antagonists and contraceptives in general was proved only for the gestagen medroxyprogesterone. There were 18 children among 130 exposures (13.85%) with malformations of all kinds. When held against the total group of uniform malformations, this statistical significance however disappears again (Nocke 1978). Apart from the synthetic DES, estrogens did not cause malformations in a statistically significant number, different from spontaneous malformations.

The suspected teratogenicity of hormonal pregnancy tests – since withdrawn from the market – could neither be proved; their intentional ingestion for abortion purposes after failing to bring about a menstrual bleeding has surely to be considered as a potential factor for embryotoxicity. Such ingestion will be difficult to exclude in the moment of attributing malformations to such products.

This is one of various critical arguments by Nocke regarding hormonal teratogenicity reports. It applies also to the intentionally or accidentally continued ingestion of contraceptives after a breakthrough pregnancy has set in.

Another fact not sufficiently taken in account was the requirement that the moment of the drug ingestion has to coincide with the embryogenetic phase of the malformed anatomical structure. If such a correlation cannot be proved a teratogenic action cannot be claimed for the product in question.

The tendency of nature to expulse a malformed conceptus by spontaneous abortion is well known. If by pharmaceutical means, i. e., exogenous sexual hormones a threatened abortion is inhibited, a malformation seen at birth will readily be attributed to this treatment. This is a frequent cause for wrong claims of teratogenicity. In the mentioned collaborative perinatal project, hemorrhages previous to such anti-abortive therapeutic measures were not sufficiently accounted for.

According to Nocke (1978), Heinonen et al. (1977) closed their book with the following statement: "The separate and combined roles of estrogenic and progestational agents need to be clarified, and it is particulary important to evaluate the effect of inadvertent use of oral contraceptives after conception."

The Heinonen report therefore offers no causal relationship that could be accepted even as a valid hypothesis for a connection between exposure to hormones during pregnancy and congenital malformations (Nocke 1978). Other large scale teratogenicity studies (Kulander and Källén, Malmoe 1976, Haller, Göttingen 1974/1975 and the prospective Jerusalem perinatal study by Harlap 1975, all discussed by Nocke) also fail to prove a conclusive evidence for the teratogenic effect of sexual hormones apart from the DES action described above and apart from the masculinization of female fetuses after the oral or intramuscular application of synthetic progestines that was described by Wilkins and that has been reproduced in female rat fetuses by Revesz (Wilkins et al. 1958; Wilkins 1960; Revesz et al. 1960).
(Barcelona, 10th of September 1979)

References

Antonioli DA, Rosen S, Bourke L, Donahue V (1979) Glandular dysplasia in diethylstilbestrol-associated vaginal adenosis. A case report and review of the literature. Am J Clin Pathol 71:715–721

Barber HRK, Sommers SC (1974) Vaginal adenosis, dysplasia and clear-cell adenocarcinoma after diethylstilbestrol treatment in pregnancy. Obstet Gynecol 43:645–652

Bibbo M, Al-Naqeeb M, Baccarini I, Gill W, Newton M, Sleeper KM, Sonek MG, Wied GL (1975) Follow-up study of male and female offspring of DES-treated mothers. A preliminary report. J Reprod Med 15:29–32

Bibbo M, Gill WB, Azizi F, Blough R, Fang VS, Rosenfield RL, Schumacher GFB, Sleeper K, Sonek MG, Wied GL (1977) Follow-up study of male and female offspring of DES-exposed mothers. Obstet Gynecol 49:1–8

Cosgrove MD, Benton B, Henderson BE (1977) Male genitourinary abnormalities and maternal diethylstilbestrol. J Urol 117:220–222

Editorial The Lancet (1974) Are sex hormones teratogenic? Lancet II: 1489–1490

Fetherston WC (1975) Squamous neoplasia of vagina related to DES-syndrome. Am J Obstet Gynecol 122:176–181

Folkman J (1971) Transplacental carcinogenesis by stilbestrol. N Engl J Med 285: 404–405

Forsberg JG (1972) Estrogen, vaginal cancer and vaginal development. Am J Obstet Gynecol 113:83–87

Frohberg H (1977) An introduction to research in teratology. Methods in prenatal toxicology. Teratology workshop April 1977, Berlin. Thieme, Stuttgart, p 1–13

Frohberg H (1978) Kanzerogenese, Teratogenese, Mutagenese. Beziehung zwischen tierexperimentellen und klinischen Befunden. Arzneim Forsch/Drug Res 28 (II) Heft 11a:1984–2001

Gill WB, Schumacher GFB, Bibbo M (1976) Structural and functional abnormalities in the sex organs of male offspring of mothers treated with DES. J Reprod Med 16: 147–153

Goldstein A (1974) Principles of drug action, Wiley, New York, p 760

Greenwald P, Barlow JJ, Nasca PC, Burnett WS (1971) Vaginal cancer after maternal treatment with synthetic estrogens. N Engl J Med 285:390–393

Harlap S, Prywes R, Davies AM (1975) Birth defects and progesterones in pregnancy. Lancet I:682–683

Hart WR, Townsend DE, Aldrich JO, Henderson BE, Roy M, Benton B (1976) Histopathologic spectrum of vaginal adenosis and related changes in stilbestrol exposed females. Cancer 37:764–775

Heinonen OP, Slone D, Shapiro S (1977) Birth defects and drugs in pregnancy. Publishing Sciences Group Littleton, Mass USA

Herbst AL, Scully RE (1970) Adenocarcinoma of the vagina in adolescence. A report of 7 cases including 6 clear-cell carcinomas (so-called mesonephromas) Cancer 25: 745–757

Herbst AL, Ulfelder H, Poskanzer DC (1971) Adenocarcinoma of the vagina; association of maternal stilbestrol therapy with tumour appearance in young women. N Engl J Med 284:870–881

Herbst AL, Kurman RJ, Scully RE, Poskanzer DC (1972) Clear-cell adenocarcinoma of the genital tract in young females. Registry report. N Engl J Med 287:1259–1264

Herbst AL, Robboy SJ, Scully RE, Poskanzer DC (1974) Clear-cell adenocarcinoma of the vagina and cervix in girls: Analysis of 170 registry cases. Am J Obstet Gynecol 119:713–724

Herbst AL, Scully RE, Robboy SJ (1975) Effects of maternal DES ingestion on the female genital tract. Hosp Pract 10:51–57

Herbst AL (1976) Summary of the changes in the human female genital tract as a consequence of maternal diethylstilbestrol therapy. J Toxicol Environ Health (Suppl) 1:13–20

Ivankovic S (1975) Praenatale Carcinogenese. In: Grundmann (ed) Geschwülste. Tumors III. Springer, Berlin Heidelberg New York (Handbuch der allgemeinen Pathologie, Vol 6, p 942–1002)

Janerich DT (1975) The pill and subsequent pregnancies. Lancet I:681–682

Janerich DT, Piper JM, Glebatis DM (1974) Oral contraceptives and congenital limb reduction defects. N Engl J Med 291:697–700

Kurman RJ (1979) Abnormalities of the genital tract following stilboestrol exposure in utero. In: Lingeman CH (ed) Carcinogenic hormones. Springer, Berlin Heidelberg New York (Recent results in cancer research, Vol 66, p 161)

Langmuir AD (1971) New environmental factor in congenital disease. N Engl J Med 284:912–913

McLachlan JA, Newbold RR (1975) Reproductive tract lesions in male mice exposed prenatally to DES. Science 190:991–992

Monaghan JM, Sirisena LAW (1978) Stilboestrol and vaginal clear-cell adenocarcinoma syndrome. Brit med J 1:1588–1590

Nocke W (1978) Sind weibliche Sexualsteroide teratogen? Gynaekologe 11:119–141

Nora JJ, Nora AH (1973) Birth defects and oral contraceptives. Lancet I:941–942

Preuschen H von (1877) Ueber Cystenbildung in der Vagina. Virchows Arch Pathol Anat 70:111

Revesz C, Chappel CI, Gandry R (1960) Masculinisation of female fetuses in the rat by progestational compounds. Endocrinology 66:140–143

Smith OW (1948) Diethylstilbestrol in the prevention and treatment of complications of pregnancy. Am J Obstet Gynecol 56:821–834

Ulfelder H (1976) The stilbestrol – adenosis – carcinoma syndrome. Cancer 38:426–431

Ulfelder H (1976 a) DES – transplacental teratogen – and possibly also carcinogen. Teratology 13:101–104

Wilkins L (1960) Masculinization of female fetus due to use of orally given progestins. JAMA 118:1028–1032

Wilkins L, Jones HW, Holman GH, Stempfel RS (1958) Masculinisation of the female fetus associated with administration of oral and intramuscular progestins during gestation: Non-adrenal pseudohermaphroditism. J Clin Endocrinol Metab 18: 559–585

Uterine Tumors in Nonhuman Primates After Estrogen Exposure

C.E. Graham, H.M. McClure and D.C. Collins

Introduction

The evidence that estrogens may have an etiologic role in various human tumors has been reviewed by Hertz (1976). The evidence is not conclusive, therefore it is necessary to turn to animal studies to obtain direct evidence that estrogens can be carcinogenic.

Estrogen administration replicably induces tumors in a variety of mammalian species and organ sites. These include tumors of breast, cervix, uterus, ovary, pituitary, and kidney in mice, rats, rabbits, hamsters, and dogs (Table 1.). The repeated failure of similar experiments to induce neoplasms in monkeys brings into question the significance of the findings from lower mammals for man (Table 2.).

The present series of studies will show that chronic diethylstilbestrol (DES) or estradiol benzoate (EB) treatment can replicably induce tumors at high frequency in a South American primate, the squirrel monkey *(Saimiri sciurcus).*

Table 1. Estrogen induced tumors in lower mammals: Selected examples

Species	Site	Compound	Reference
Mouse	Breast	Estradiol Mestranol	Rudali et al. 1971
Mouse	Leydig cell	Triphenylethylene, tri-p-anisyl- -chloroethelene	Shimkin et al. 1941
Mouse	Uterine serosa and cervix	EB DES	Gardner et al. 1956
Rat	Adrenal	Estrone	Noble, 1967
Rat	Pituitary	E_2 valerate	Sonnenschein et al. 1974
Hamster	Kidney	DES	Lacomba and Gabaldon 1971
Rabbit	Uterus, serosa	DES	Meissner et al. 1957
Dog	Ovary, serosa	DES	Jabara 1959
Dog	Uterine serosa	"Synthetic E"	Mawdesley et al. 1968
Dog	Serosa	Trans-4, 4^1-dimethyl-α, α- -diethylstilbene	Owen et al. 1972

Table 2. Chronic estrogen treatment of primates

Reference	Drug	Dose	Time	dose	status	n
Engle and Smith 1935	E_1	84–168[a] μg/day	60–110 d	9.8–40 mg	Adult oox Rh+	4
Hisaw and Lendrum 1936	E_1	17–84[a] μg	22–103 d	0.45–3.4 mg	Young Rh	4
Overholser and Allen 1935	E_1	0.02–151 μg/day[a]	81–90 d	0.69–2 mg	Adult oox Rh	4
Scott and Wharton 1955	DES	0.8 mg/day i.m.	35–24 mon.	0.39–36.5g	Rh	4
Engle et al. 1943	E_2	50 mg/5–6 wk	24–27 mon.	575–825 mg	Aged Rh, 2/5 oox	5
Iglesias and Lipschutz 1947	E_2	S.C.	203–254 d	41.5–90.1 mg	Cebus	2
Zuckerman 1937	E_2 1	100 μg/	365 d	36.5 mg	Rh	1
Dahl-Iverson et al. 1942	EB	10–100 μg/day i.m.	15 mon.	28–37 mg	Young Rh	2
Dahl-Iverson et al. 1942	DES	100 μg/day i.m.	15 mon.	28 mg	Adult Rh	1
Hartman et al. 1941	Various	>3–12 mg/wk	⩽27 mon.	–	Young Rh, f.	20

References:
(1) Engle & Smith, 1935; (2) Hisaw & Lendrum, 1936; (3) Overholzer & Allen, 1935; (4) Scott & Wharton, 1955; (5) Engle et al, 1943; (6) Iglesias & Lipschutz, 1947; (7) Zuckerman, 1937; (8) Dahl-Iverson et al, 1942; (9) Hartman et al, 1941.

a Bioassay values expressed as equivalent activity of estrone.

\+ Rh: rhesus; f: female

The tumors induced in three separate studies were derived from the peritoneal epithelium (serosa) and usually originated from the uterine corpus. Similar lesions sometimes occur in humans and are called pelvic or peritoneal mesotheliomas or adenomatoid tumors.

Methods

Squirrel monkeys were implanted subcutaneously with four cylindrical pellets of hormone of total weight 250 mg. When pellets were lost by ulceration through the skin or dissolution, they were replaced.
The animals were caged in groups of 5–10, and were fed monkey chow, apples,

and oranges, and had free access to water. After death a thorough post-mortum was done and representative portions of all major organs were fixed in Bouins fluid or neutral buffered formalin: 6μ sections were stained with hematoxylin and eosin.

Results

The first study was described in detail by McClure and Graham in 1973; four control animals imported from Columbia and Peru were implanted with cholesterol. Ten animals were implanted with pure DES, and killed 5–14 months later.

No neoplastic lesions were found in the controls, but three DES-treated animals had serosal hyperplasia and seven had invasive serosal lesions. (Table 3.)

The normal serosa is a delicate squamous epithelium (Fig. 1). The DES-treated animals all showed varying degrees of uterine hypertrophy and endometrial hyperplasia. Microscopic examination revealed serosal lesions in all DES treated animals.

Table 3. Incidence of serosal lesions in colombian/peruvian squirrel monkeys (group I)

	Number of subjects	Number with serosal hyperplasia	Number with invasive mesothelioma
Controls	4	0	0
DES treated (5–14 months)	10	3	7 (2)[a]

[a] 2 of these animals had extrauterine foci of serosal neoplasia

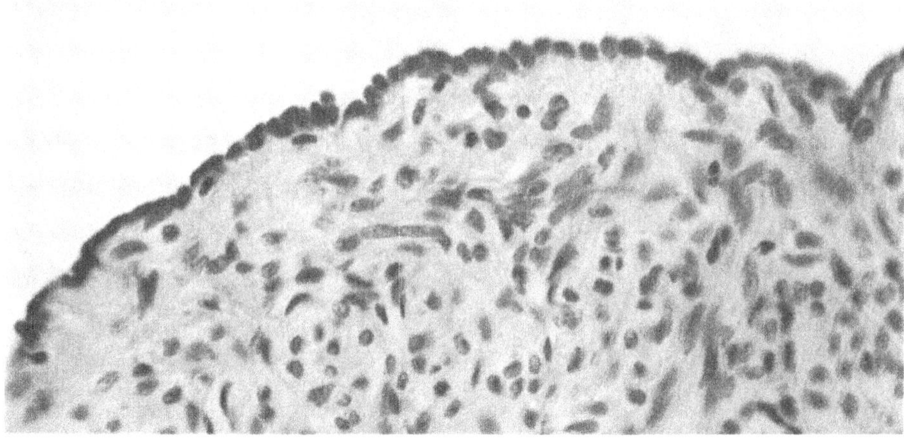

Fig. 1. Normal uterine serosa of squirrel monkey. H and E stain. x 400

The least severely affected had a hypertrophic cuboidal or columnar uterine serosa, aggregated into several layers. Elsewhere there was superficial penetration into the myometrium, or papillary proliifferations. In several cases there was invasion of the myometrium by solid sheets of cells or ductlike structures that formed pseudoglandular cavities in the outer myometrium and papillary formations (Fig. 2).

The most severely affected uteri had similar, though much more extensive changes with numerous or abnormal mitotic figures and widespread infiltration of the endometrium. In one instance, the infiltrating serosal cells reached the endometrium, but remained distinguishable from it by a lack of apical cytoplasm, lesser nuclear staining and a disorganized pseudoglandular pattern.

One animal had grossly visible uterine outgrowths (Fig. 3) composed of sheets of cells and acinar formations, (Fig. 4) numerous nodules spread throughout the mesentery, and similar lesions on the surface of the adrenal and spleen. The extrauterine lesions cytologically resembled the uterine serosal neoplasms.

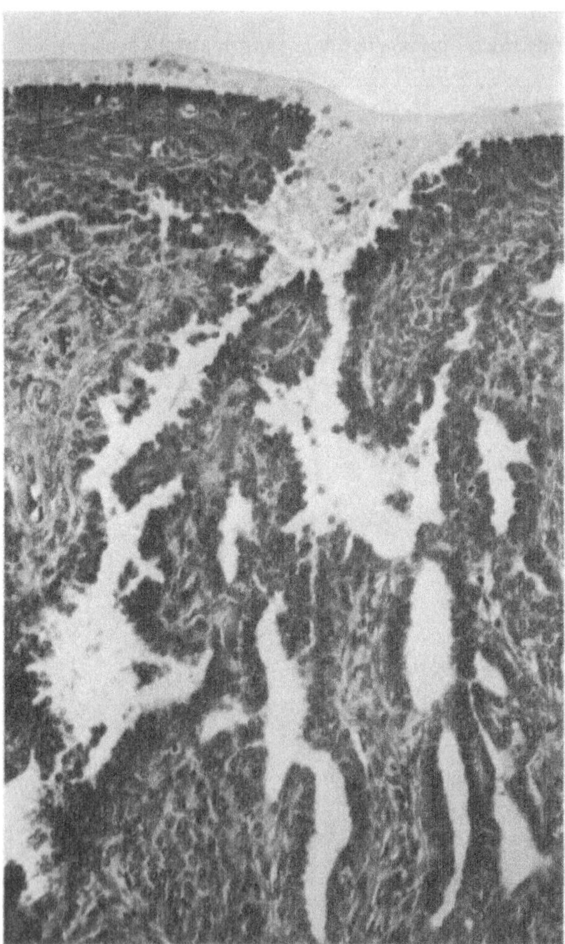

Fig. 2. Proliferation of uterine serosa of DES-treated squirrel monkey, showing papillary formations and formation of pseudoglandular spaces. H and E stain. x 200

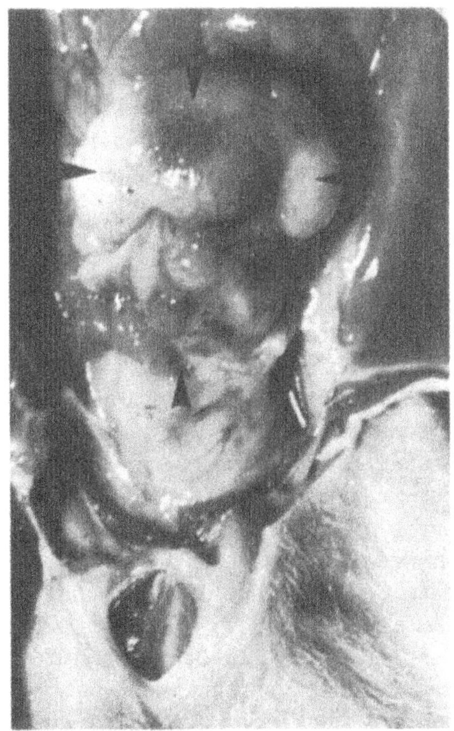

Fig. 3. Gross appearance of greatly enlarged uterus, 14 months after initiation of DES treatment; the extent of the uterus is delineated by arrows (note the irregular outgrowths covering its surface) x 1

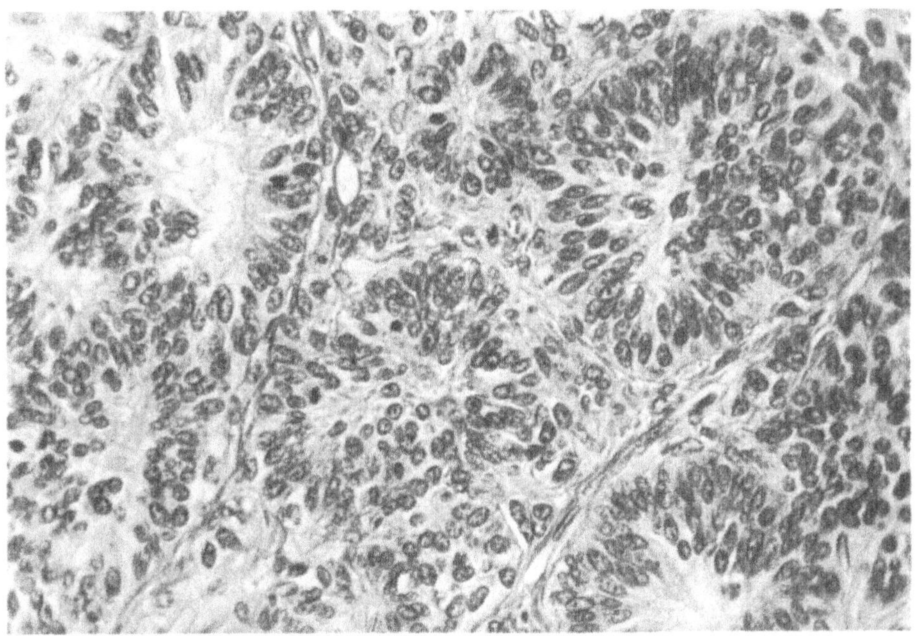

Fig. 4. Microscopic structure of area of outgrowth, showing cells of serosal origin packed into an acinar pattern and prominent mitotic figures. H and E stain. x 400

The degree of development of the tumors in these ten DES-treated animals was correlated with the duration of DES exposure. The series of stages from serosal hyperplasia, superficial and deep myometrial invasion, and finally endometrial penetration, left no doubt of the serosal origin of all the lesions and permitted a confident diagnosis of mesothelioma. Based upon their invasive pattern, and in some cases presence of numerous or abnormal mitotic figures, seven of the ten mesothelial lesions were considered to be malignant.

Two additional studies were since undertaken to verify these findings and to investigate the malignant potential and biologic properties of the estrogen induced mesotheliomas in more detail.

The second group of monkeys comprised three untreated controls, nine DES-treated for up to 28 months and one EB treated for 28 months. (Table 4.). Initially the DES treated animals received 1, 4, or 16 pellets weighing 60 mg. After 1 year these pellets could not be palpated and all animals were reimplanted with four pellets each. During monthly examinations, pellets were found to be lost with increasing frequency, and so 21 months after the first implant, weekly injections of 5 mg DES in oil were also administered, and continued for 2–6 months. Thereafter the animals were examined frequently until death. Animals that survived for more than 28 months were monitored for 40–54 months before being killed. Consistent with the longer period of estrogen exposure, more advanced tumors were found.

Three of four DES-treated animals that died or were killed 22–28 months after commencing DES exposure had greatly enlarged uteri with prominent uterine outgrowths, and extra-uterine foci of serosal neoplasia. Uterine lesions often showed nuclear hyperchromasia and pleomorphism, packing of nuclei and frequent mitoses.

Extrauterine lesions were found in omental mesenteric lymph nodes, and on the surface of liver, spleen, pancreas, and adrenal gland. In most cases it was uncertain whether these were separate loci of neoplasia, had originated by seeding from the original peritoneal location or neoplasia, or were metastases spread through the circu-

Table 4. Incidence of serosal lesions in columbian/peruvian squirrel monkeys (group II)

	Number of subjects	Number with serosal hyperplasia	Number with invasive mesothelioma
Controls	3	0	0
DES treated (22–28 months)	9	0	4 (3)[a]
EB treated (28 months)	1	0	1 (1)[a]

[a] Extrauterine foci of serosal neoplasia

latory system. In one animal definite metastases were found in axillary lymph node, sublumbar muscle and subcutaneously in the umbilical region.

One additional animal killed 25 months after initiation of DES exposure had a mesothelioma restricted to the myometrium.

A single EB-exposed animal had prominent uterine outgrowths, myometrial infiltration, numerous nodules in the omentum, and serosal proliferations on the surface of the colon, mesenteric lymph nodes, liver, bladder, spleen; tumor cells largely replaced the sublumbar lymph nodes.

The EB- and one DES-induced tumor was grown in tissue culture for five passages. The cells showed loss of contact inhibition, a characteristic of tumor cells.

The EB-induced tumor culture was subjected to DNA analysis after ethanol fixation, acid hydrolysis, and feulgen staining. Using squirrel monkey sperm and human lymphocytes as 1 and 2n standards, and measuring the absorption of 100 nuclei, the ploidy of the tumor cells was estimated: 6% were diploid, 26% tetraploid, and 70% were aneuploid.

The noncapsulated, proliferative, invasive, multifocal, and metastic properties of these five mesotheliomas, combined with their cytologic characteristics of nuclear pleomorphism, hyperchromasia, and aneuploidy provide compelling evidence that these tumors are malignant.

A third, larger group of animals were of Bolivian origin (Table 5). Of six untreated controls, one died after 35 months (hypothermia) and five were killed after 62 months. No tumors were found in this group.

Five animals were implanted with EB. Two died after 32–36 months of peritonitis, three were killed after 42 months. One of these had mesothelioma invading the myometrium but no extrauterine neoplasia. All showed uterine hypertrophy and hyperplasia. two had extreme adenomeous hyperplasia.

Twenty-two animals were implanted with DES. This group of animals tolerated pellets poorly and consequently after 12–16 months the animals were supplemented with weekly 5 mg DES injections, up to a total of 16–17 months DES exposure.

Of these 22 DES treated animals, 13 died or were killed as a result of poor health

Table 5. Incidence of serosal lesions in bolivian squirrel monkeys (group III)

	Number of subjects	Number with serosal hyperplasia	Number with invasive mesothelioma
Controls	6	0	0
DES treated (10–17 months)	22	1	9 (1)[a]
EB treated (42 months)	5	0	1

[a] One additional animal had mesothelioma of oviduct

after 10–27 months treatment. Gross findings were emaciation, dehydration, peritonitis, and nephritis.

Myometrial mesothelioma was found in 6 out of 13 animals (46%): most lesions were microscopic. In three out of nine animals that were killed 40–42 months after initiation of DES, and 23–28 months after DES withdrawal, regressed lesions were found in the myometrium.

One additional animal had extensive invasion of serosal cells into the oviductal wall. Thus a total of ten mesotheliomas were found in this group of animals. No multifocal lesions were found in this group. The presence of regressed lesions in three animals that survived withdrawal of DES by 2 years or more suggests that continued tumor development requires continued estrogen exposure: in other words there is no evidence that tumors develop to estrogen independence.

Conclusions

Table 6 summarizes the overall tumor incidence in the three groups. Spontaneous mesothelioma has never been observed in approximately 300 squirrel monkeys examined by us, including the 13 controls assigned to this study. Twenty-one malignant uterine mesotheliomas in 41 DES-treated animals and two in six EB-treated animals are documented here, a convincing case for carcinogenicity of estrogens in a non-human primate.

The observation that estrogen treatment can induce neoplasms in adults of one primate species, as well as in lower mammals, reinforceds the widespread suspicion that estrogens have carcinogenic potential in adult humans also. It is interesting to note that estrogens can induce serosal neoplasias in a variety of species (Table 1), although no association between estrogen exposure and human pelvic mesothelioma is known.

In squirrel monkeys, similar lesions were induced with the nonsteroidal estrogen, DES, and the steroid estrogen EB. Other estrogens have induced serosal and other tumors in lower mammals.

The fact that a variety of different estrogenic molecules (Table 1) can produce tumors, often histologically similar, suggests that their carcinogenicity is a function of a property that these substances share in common, namely estrogenicity. Thus the

Table 6. Incidence of uterine serosal lesions in squirrel monkeys (summary – groups I – III)

	Number of subjects	Number with serosal hyperplasia	Number with invasive mesothelioma (%)
Controls	13	0	0
DES treated	41	4	21 (50%)
EB treated	6	0	2 (33%)

hypothesis that carcinogenicity is a general property of potent estrogens could be tested in squirrel monkeys, and data could be obtained in this species on the carcinogenic potential of estrogenic compounds destined for human use.

The existance of serosal lesions in a variety of animal species after estrogen treatment suggests that an association with prior estrogen exposure should be sought in cases of human pelvic mesothelioma.

Acknowledgments

This work supported by DHEW-NIH grants CA /15656 from the National Cancer Institute and RR /00165 from the Animal Resources Branch. We thank Dr. J. Chang and Z. Olkowski for performing tissue culture and cytospectrophotometric analysis.

References

Dahl-Iversen E, Hamburger C, Jørgensen H (1942) Cystic glandular hyperplasia of the endometrium elucidated by therapeutic experiences in patients and by experiments on rhesus monkeys. Acta Obstet Gynecol Scand 5:315

Engle ET, Smith PE (1935) Some uterine effects obtained in female monkeys during continued oestrin administration, with especial reference to cervix uteri. Anat Rec 61:471

Engle ET, Krakower C, Haagensen CD (1943) Estrogen administration to aged female monkeys with no resultant tumors. Cancer Res 3:858

Gardner WU, Ferrigno M (1956) Unusual neoplastic lesions of the uterine horns of estrogen-treated mice. J Natl Cancer Inst 17:601

Hartman CG, Geschickter CF, Speert H (1941) Effects of continuous estrogen administration in very large doses. Anat Rec (Suppl 2) 79:31

Hertz R (1976) Steroid-induced, steroid-producing, and steroid-responsive tumors. In: Menon KMJ, Reel JR (eds) Steroid hormone action and cancer. Plenum, New York, pp 1–14

Hisaw FL, Lendrum FC (1936) Squamous metaplasia in the cervical glands of the monkey following oestrin administration 1,2. Endocrinology 20:228

Iglesias R, Lipschutz A (1947) Effects of prolonged oestrogen administration in female New World monkeys with observations on a pericardial neoplasm. Endocrinol 5:88

Jabara AG (1959) Canine ovarian tumours following stilboestrol administration. Aust J Exp Biol Med Sci 37:549

Lacomba T, Gabaldon M (1971) Biochemical studies of diethylstilbestrol-induced kidney tumors in the Golden Syrian hamster. Cancer Res 31:1251

Mawdesley-Thomas LE, Sortwell RJ (1968) Proliferative lesions of the canine uterus associated with high dose desterogen administration. Vet Rec 82:468

McClure HM, Graham CE (1973) Malignant uterine mesotheliomas in squirrel monkeys following diethylstilbestrol administration. Lab Anim Sci 23:493

Meissner WA, Sommers SC, Sherman G (1957) Endometrial hyperplasia, endometrial carcinoma, and endometriosis produced experimentally by estrogen. Cancer 10:500

Noble RL (1967) Induced transplantable estrogen-dependent carcinoma of the adrenal cortex in rats. Proc Am Assoc Cancer Res 8:51 (Abst /200)

Overholser MD, Allen E (1935) Atypical growth induced in cervical epithelium of the monkey by prolonged injections of ovarian hormone combined with chronic trauma. Surg Gynecol Obstet 60:129

Owen NV, Pierce EC, Anderson RC (1972) Papillomatous growths on internal genitalia of bitches administered the synthetic estrogen trans-4, 4-dimethyl-, -diethylstilbene. Toxicol Appl pharmacol 21:582

Rudali G, Coezy E, Frederic F, Apiou F (1971) Susceptibility of mice of different strains to the mammary carcinogenic action of natural and synthetic oestrogens. Eur J Clin & Biol Res 16:425

Scott RB, Wharton LR Jr, (1955) The effects of excessive amounts of diethylstilbestrol on experimental endometriosis in monkeys. Am J Obstet Gynecol 69:573

Shimkin MB, Grady HG, Andervont HB (1941) Induction of testicular tumors and other effects of stilbestrolcholesterol pellets in strain C mice. J Natl Cancer Inst 2:65

Sonnnenschein C, Posner M, Sahr K, Farookhi R, Brunelle R (1974) Estrogen sensitive cell lines. Establishment and characterization of new cell lines from estrogen-induced rat pituitary tumors. Exp Cell Res 84:399

Zuckerman S (1937) Effects of prolonged oestrin-stimulation on the cervix uteri. Lancet 232:435

Morphological Changes Induced in the Human Uterus and Fallopian Tube by Exogenous Estrogens

Gisela Dallenbach-Hellweg

Abstract

Although the estrogens do show fine distinctions in their effects that are of clinical importance, most of these hormones are extremely potent, even in tiny doses, since the target cells they stimulate are exquisitely responsive to low concentrations and promptly react by characteristic changes. On the other hand, prolonged high doses seem to overwhelm the receptors of the target cells and apparently exhaust the cytoplasmic structures that produce them so that specific enzyme systems become blocked, and the cells become unable to respond to further estrogen. Atrophy ensues.

The administration of estrogen alone from the first day of a menstrual cycle prolongs the proliferative phase and suppresses the secretion of gonadotropins by the pituitary, especially FSH. Development of a corpus luteum is inhibited until the estrogen is discontinued. Menstruation consequently is delayed. Treatment with estrogen during the secretory phase results in severe stromal edema, delay in secretory transformation of glands and stroma and a disruption of the nucleolar channel-system. If the estrogen is discontinued after prolonged therapy, an estrogen "withdrawal bleeding" ensues; if the estrogen is given over a long period in consistently small doses, a spontaneous "breakthrough bleeding" occurs. With prolonged administration of small doses of estrogen (20–100 μg daily) the endometrium responds with glandular-cystic hyperplasia. Only estriol is less harmful to the endometrium. From time to time portions of these hyperplastic endometria may undergo hemorrhagic necrosis and be discharged, but when the estrogen stimulus continues, unopposed by progesterone, they may progress to precancerous adenomatous hyperplasias. Droplets of fat appear in stromal cells of the upper layers of the endometrium and foam cells eventually develop. In contrast to the hyperplasias arising in the pre- and postclimacteric periods because of endogenous causes, those developing after many years of estrogen therapy are characterized by special morphological features: they develop multicentrically and even in the adenomatous stage may be circumscribed or polypoid. In the adenomatous regions the appearance of the nuclei and cytoplasm of the epithelial cells varies from gland to gland. Both flattened and nodular formations of metaplastic squamous epithelium are especially common.

Whereas the endocervix shows no effect, the ectocervical epithelium reacts with hyperproliferation and keratinization. When estrogen is given to neonatal mice, these changes as well as squamous metaplasia and dysplasia may persist throughout life.

The epithelium of the fallopian tube is also induced to proliferate atypically.

In all known target organs in both humans and animals, estrogens bind to specific receptors, form a receptor-estrogen complex in the cell nucleus, and thereby activate DNA synthesis. That results in RNA transcription and increased protein production with the accompanying processes of growth and proliferation. A wave of mitosis ensues.

The estrogens used in therapy consist of various natural steroids and nonsteroid, synthetic substances. Of the naturally occurring estrogens, estradiol, estrone, and estriol, estradiol has the greatest affinity to the estrogen receptors. The substances most commonly used are orally effective derivates of 17-α-ethinyl-estradiol or its methylether, mestranol, in addition to conjugated estrogens.

When administered alone at the beginning of the cycle, estrogen will lengthen the proliferation phase, delay ovulation and menstruation, and finally lead to withdrawal bleeding similar to that seen in follicle persistence. The characteristic histologic picture is one of hemorrhagic necroses in an endometrium that is irregularly proliferated. If continued, small to medium doses of estrogen will cause hyperplasias, the severity of which depends on length of medication and which run from glandular-cystic to precancerous adenomatous forms (Fig. 1; cf. Dallenbach-Hellweg 1964; Manning et al. 1971).

Hyperplasias of these types have been induced in a number of species and with almost all derivates of estradiol. Similar effects have been known to occur in humans for some time (Schroeder 1954; Bloomfield 1957; Greenblatt and Zarate 1967; Ober and Bronstein 1967), and are reproducible in tissue culture (Csermely et al. 1971). Estriol given alone seems to have little effect on the corpus endometrium (Sjöstedt and Strandh 1971). The hyperplasias that appear after long-term estrogen therapy have particular morphological characteristics which distinguish them from endogenous hyperplasias, both pre- and postmenopause. They begin multicentrically, and even when they have reached the adenomatous stage may still be focal and often polypous. The structure of these adenomatous overgrowths may differ from gland to gland, depending on the appearance of nucleus and cytoplasm. Flat and nodular metaplasias of the squamous epithelium are particularly common. Droplets of lipid appear in the cells of the upper stromal layers, and groups of steroid-containing foam cells are also observed.

Reports are becoming more and more frequent of the carcinomatous transformation of endometrium following estrogen-induced hyperproliferation. A number of authors have described the clinical development of endometrial carcinoma following long-term estrogen medication, some of them having observed all the stages of hyperplasia leading up to the carcinoma (Corscaden and Gusberg 1947; Novak and Rutledge

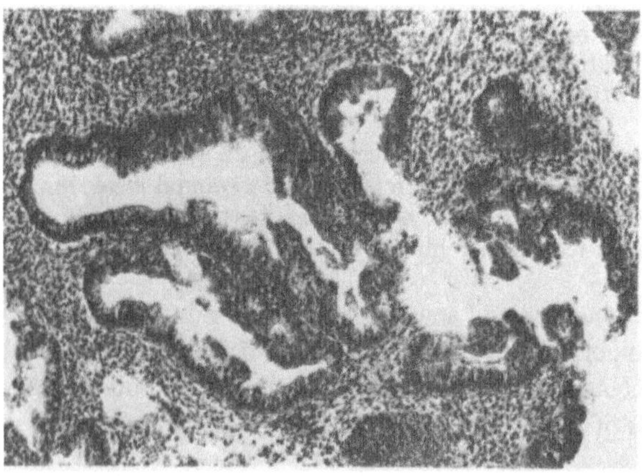

Fig. 1. Adenomatous hyperplasia of the corpus endometrium after estrogen administration. H and E stain. x 96.1

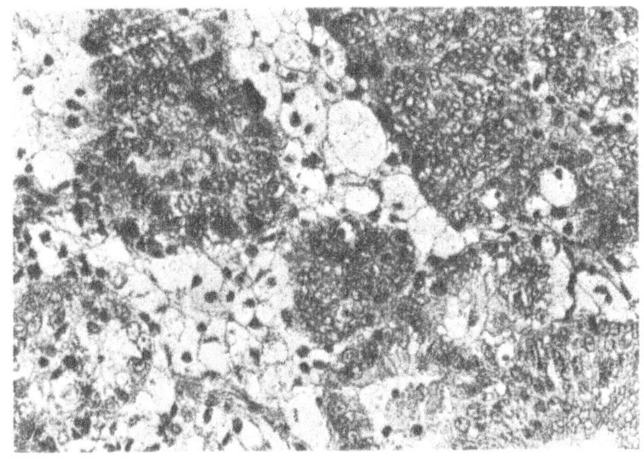

Fig. 2. Papillary adenocarcinoma of the corpus endometrium following long-term estrogen therapy. Note large numbers of endometrial foam cells in the remaining stromal cunea. H and E stain. x 279.5

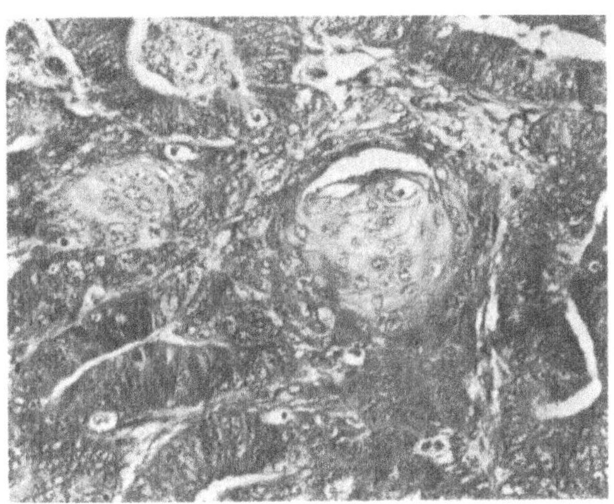

Fig. 3. Adenoacanthoma of the corpus endometrium after several years of estrogen therapy, showing metaplasia of squamous epithelium in glandular epithelium. H and E stain. x 235.2

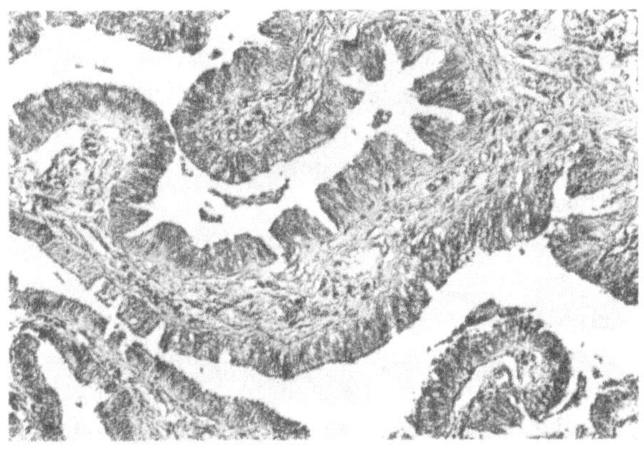

Fig. 4. Highly proliferated, partially multilayered tube epithelium after estrogen administration, showing adenomatous hyperplasia of the endometrium. H and E stain. x 121 (Dallenbach-Hellweg and Niehoff 1971)

1948; Speert 1948; Jensen and Ostergaard 1954; Wallach and Hennemann 1959; Boutselis et al. 1963; Hertz 1967; Laufer 1968 and many more). Riehm and Stoll (1952) noted the histologic particularities of these carcinomas, with their numerous epithelial papillae extending into heavily branched glandular lumina, their multicentric induction, and their relatively high degree of maturation. Gusberg and Hall (1961) considered the adenomatous hyperplasias and carcinomas that appeared following estrogen therapy to have such a characteristic glandular picture, they went so far as to dub them "estrogen carcinomas" (Fig. 2). From our own material, which grows by the day, we can confirm their observations that these carcinomas show characteristics with kaleidoscopic variation in glandular shapes and a tendency for metaplasia (Fig. 3).

The epithelium of the *fallopian tube* reacts to estrogen given alone with atypical proliferation, multiple layer formation, and the formation of intraluminal epithelial papillae (Fig. 4; Dallenbach-Hellweg and Niehoff 1971). The extent of proliferation increases with length of administration, though it remains somewhat behind that of the endometrium, apparently because the epithelium of the fallopian tube is somewhat less susceptible to estrogen stimulus than the endometrium.

The latest information that has become available on the mechanism of effect of estrogen and its close involvement with gene activity and cell division, has lent further support to the suspicion that estrogen may indeed be a cocarcinogen if nothing more. The most recent American case-control-studies, which included 451 carcinomas and 888 controls, have revealed that women who had had estrogen therapy stood a 6 to 15 times greater risk of contracting endometrial carcinoma than those who had not (Antunes et al. 1979). Another study has revealed that not only length of medication but size of dose may play an important role in the induction of cancer (Gray et al. 1977). Still another investigation has shown that of 94 postmenopausal patients with endometrial carcinoma, 70% had taken estrogens compared to only 23% of a control group of same-age women without cancer (Ziel and Finkle 1976). By the same token, a drastic reduction in estrogen medication led to considerably lower risk of illness (Jick et al. 1979).

In recent years, a higher frequency of adenomatous hyperplasia and endometrial carcinoma has been observed in young women who had taken contraceptives high in estrogen (Lyon 1975; Silverberg and Makowski 1975; Kelley et al. 1976; Cohen and Deppe 1977; Reeves and Kaufman 1977; Silverberg et al. 1977). This has been explained by the relative predominance of estrogen during continual, artificially induced anovulation.

When estrogen therapy is discontinued, even a high-grade adenomatous hyperplasia may spontaneously regress; the endogenous form, however, requires high doses of gestagens to induce successful regression. These exogenous estrogen-related disorders become irreversible only when they reach the stage of manifest carcinoma; and before menopause, only when the myometrium has been invaded.

Toxically high doses of estrogen have led, both in animals and humans, to atrophy of the endometrium (Hartman et al. 1941).

The mucus membrane of the *cervix* reacts in a basically different manner to endocrine stimulation than does the corpus endometrium. Though it increases mitosis in the endometrium, estradiol administered to neonatal mice tends to decrease mitosis in the endocervix (Forsberg 1971). Initially the human endocervix does not react at all to estrogen, though the surface epithelium may evince bizarre nuclei. With continued administration, atrophy ensues, and the border between squamous and cylindric epithelium is displaced. Differentiation is another story; the cervix glands produce large amounts of watery secretion consisting of carbohydrate-rich glycoproteins of the mucoid type of which only a very small proportion are soluble (Schumacher 1970).

Short-term estrogen administration leads to proliferation of the ectocervical epithelium to up to 40 layers (Bo 1970). The S-phase of DNA synthesis is measurably shortened under these conditions (Galand et al. 1971). Chronic estrogen administration leads, above and beyond this, to the development of an extensive parakeratosis and thence to leukoplakia, and to squamous epithelial metaplasias which, during the formation of the so-called third mucus membrane, affect the endocervix in varying degrees. Histochemically, what we see in addition to a general rise in enzyme reactions, is a maturing upper layer of metaplastic epithelium that is rich in glycogens, and an immature basal layer that is sharply differentiated from the surface layer (Goslar et al. 1971).

The *myometrium*, too, reacts to estrogen stimulation with RNA and protein synthesis. In guinea pigs, estrogen has been reported to facilitate the development of fibromyomas, and in hamsters that of leiomyosarcomas. Rapid growth of hypercellular fibromyomas has also been observed clinically in patients who are on estrogen therapy.

References

Antunes CM, Stolley PD, Rosenshein NB, Davies JD, Tonascia JA, Brown C, Burnett L, Rutledge A, Pokempner M, Garcia R (1979) Endometrial cancer and estrogen use. N Engl J Med 300:9

Bloomfield A (1957) Two cases of excessive uterine hypertrophy following on prolonged oestrogen administration. J Obstet Gynaecol Br Emp 64:413

Bo WJ (1970) The effect of progesterone and progesterone-estrogen on the glycogen deposition in the vagina of the squirrel monkey. Am J Obstet Gynecol 107:524

Boutselis JG, Deneef JC, Ullery JC, George OT (1963) Histochemical and cytologic observations in the normal human endometrium. I. Histochemical observations in the normal human endometrium. Obstet Gynecol 21:423

Cohen CJ, Deppe G (1977) Endometrial carcinoma and oral contraceptive agents. Obstet Gynecol 49:390

Corscaden JA, Gusberg SB (1947) The background of cancer of the corpus. Am J. Obstet Gynecol 53:419

Csermely T, Hughes EC, Demers LM (1971) Effect of oral contraceptives on human endometrium in culture. Am J Obstet Gynecol 109:1066

Dallenbach-Hellweg G (1964) Das Karzinom des Endometrium und seine Vorstufen. Verh Dtsch Ges Pathol 48:81

Dallenbach-Hellweg G, Niehoff B (1971) Das Tubenepithel in Korrelation zu histologischen Befunden an Endometrium und Ovar. Virchows Arch [Pathol Anat] 354:66

Forsberg JG (1978) A difference in mitotic rate response between the uterine and cervical epithelium in neonatal mice after oestradiol treatment. In: Hubinont PO, Leroy F, Galand P Basic actions of sex steroids on target organs. Karger, S 248–253 Basel

Galand P, Leroy F, Chretien J (1971) Effect of oestradiol on cell proliferation and histological changes in the uterus and vagina of mice. J Endocrinol 49:243

Goslar HG, Bartels C, Grigoriadis P, Hundsdörfer G (1971) Weitere enzymtopochemische Untersuchungen an hormoninduzierten Metaplasien des Epithels der Rattenvagina. Acta Histochem (Jena) 39:217

Gray LA, Christopherson WM, Hoover RN (1977) Estrogens and endometrial carcinoma. Obstet Gynecol 49:385

Greenblatt RB, Zarate A (1967) Endometrial studies following quinestrol administration. Int J Fertil 12:187

Gusberg SB, Hall RE (1961) Precursors of corpus cancer. III. The appearance of cancer of the endometrium in estrogenically conditioned patients. Obstet Gynecol 17:397

Hartmann CG, Geschickter GF, Speert H (1941) Effects of continuous estrogen administration in very large doses. Anat Rec [Suppl 2] 79:31

Hertz R (1967) The role of steroid hormones in the etiology and pathogenesis of cancer. Am J Obstet Gynecol 98:1013

Jensen EV, Ostergaard E (1954) Clinical studies concerning the relationship of estrogens to the development of cancer of the corpus uteri. Am J Obstet Gynecol 67:1094

Jick H, Watkins RN, Hunter JR, Dinan BJ, Madsen S, Rothman KJ, Walker AM (1979) Replacement estrogens and endometrial cancer. N Engl J Med 300:218

Kelley HW, Miles PA, Buster JE, Scragg WH (1976) Adenocarcinoma of the endometrium in women taking oral contraceptives. Obstet Gynecol 47:200

Laufer A (1968) The influence of steroids on the endometrium. In J Fertil 13:373

Lyon FA (1975) The development of adenocarcinoma of the endometrium in young women receiving long-term sequential oral contraception. Am J Obstet Gynecol 123:299

Manning JP, Tornaben JA, Schwartz E (1971) Influence of quinestrol on the histology and phosphatase activity of the simian uterus. Fertil Steril 22:194

Novak E, Rutledge F (1948) Atypical endometrial hyperplasia simulating adeno-carcinoma. Am J Obstet Gynecol 55:46

Ober WB, Bronstein SB (1967) Endometrial morphology following oral administration of quinestrol. Int J Fertil 12:210

Reeves KO, Kaufman RH (1977) Exogenous estrogens and endometrial carcinoma. J Reprod Med 18:197

Riehm H, Stoll P (1952) Korpuskarzinom nach 17jähriger Follikelhormon-Medikation. Geburtshilfe Frauenheilkd 12:985

Schroeder R (1954) Endometrial hyperplasia in relation to genital function. Am J Obstet Gynecol 68:294

Schumacher CFB (1970) Biochemistry of cervical mucus. Fertil Steril 21:697

Silverberg SG, Makowski EL (1975) Endometrial carcinoma in young women taking oral contraceptive agents. Obstet Gynecol 46:503

Silverberg SG, Makowski EL, Roche WD (1977) Endometrial carcinoma in women under 40 years of age. Cancer 39:592

Sjöstedt S, Strandh J (1971) Effect of polyestriol phosphate on the vaginal cytology and uterine endometrium of postmenopausal women. Acta Obstet Gynecol Scand 50:30

Speert H (1948) Corpus cancer. Clinical, pathological, and etiological aspects. Cancer 1:584

Wallach S, Henneman PH (1959) Prolonged estrogen therapy in postmenopausal women. J Am Med Assoc 171:997

Ziel HK, Finkle WD (1976) Association of estrone with the development of endometrial carcinoma. Am J Obstet Gynecol 124:735

Effects of Hormones on Surface Structure of Human Endometrium*

R. Herbst and A.-M. Multier-Lajous

Abstract

The surface-relief of the endometrium depends on the status of the sex hormones. During the proliferative phase of a normal cycle the number of ciliary cells increases. The outgrowth of young cilia proves that this type of cell develops anew. Single cilia indicate cessation of mitotic activity and subsequent cell differentiation. Exogenously administered hormones act on the surface structure. *Estrogens* may promote the development of cilia. In glandularcystic hyperplasia not only are the ciliated cells increased, but also the number of cilia in each cell is greater. Estrogen therapy for anovulatory cycles may lead to complete differentiation. In other cases, however, the endometrial surface manifests a certain "immaturity."

In recent years a number of investigators (Ferenczy et al. 1972; Ferenczy and Richart 1973 a, b; Ferenczy 1976, 1977; Hafez et al. 1975; Ludwig and Metzger 1976 a, b, c and 1977; and Nilsson and Nygren 1972) have used the scanning electron microscope to study the human endometrium. In addition to changes it goes through during the normal cycle, they have reported on the effects that hormones have on it, as well as on tumor formation. Our study is concerned with the influence of estrogens on the surface structure of the endometrium.

Our tissue samples were fixed supravital in glutaraldehyde or buffered formaldehyde solution and embedded in paraffin[1]. Histologic information was obtained first in all samples. The remaining material was then prepared for scanning electron microscopy using the Multier-Lajous method (1971) and critical point drying.

In one case of adenomatous hyperplasia we found small, rump-shaped projections on the endometrium surface, a sign of limited proliferation (Fig. 1). The surface of these undifferentiated cells showed in low relief, and the microvilli were short and squat (Fig. 2). This picture is similar to that of undifferentiated cells during embryonal development.

Our technique enabled us to look not only at the natural surface of the endometrium, but also into glandular openings on the section face. In a case of glandular-cystic hyperplasia the glandular epithelium was extended and proliferated. In addition to secreting cells covered with microvilli we found ciliated cells in the openings of the glands (Fig. 3a). These ciliated cells were detected only rarely in cystic glands. The

---

* Prof. Dr. med. H. Lax dedicated to his 70th birthday
Support from the Deutsche Forschungsgemeinschaft is greatfully acknowledged
1 A number of the samples were kindly provided by Prof. Kloos

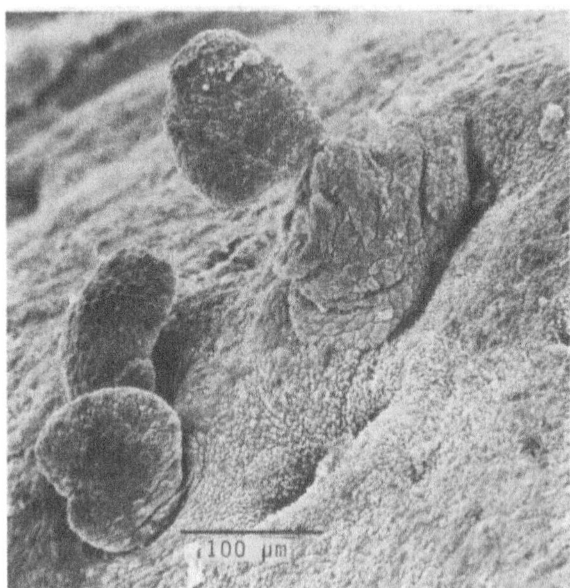

Fig. 1. Adenomatous endometrial hyperplasia showing rump-shaped protuberances on the endometrium surface

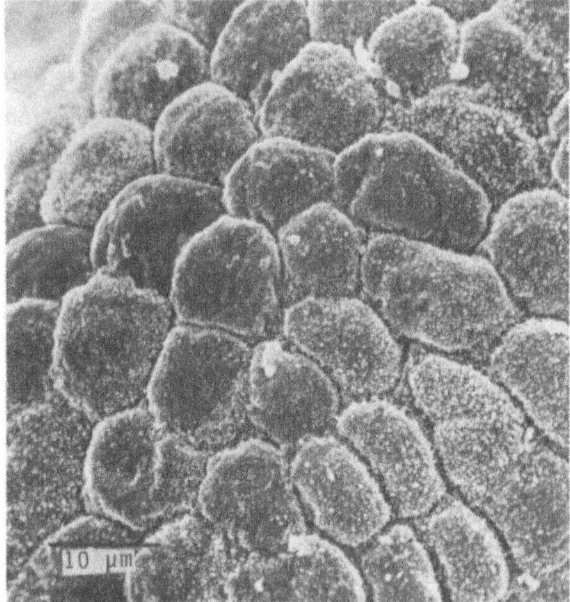

Fig. 2. Detail of Fig. 1, showing undifferentiated cells in proliferating tissue

epithelial cells were broad and flat, having the appearance almost of squamous cells (Fig. 3b).

One of our patients, who had received injections of estrogen 10 years after menopause, had developed glandularcystic hyperplasia. We found a large number of ciliated cells distributed across the epithelium of this patient (Fig. 4). The cilia were arranged

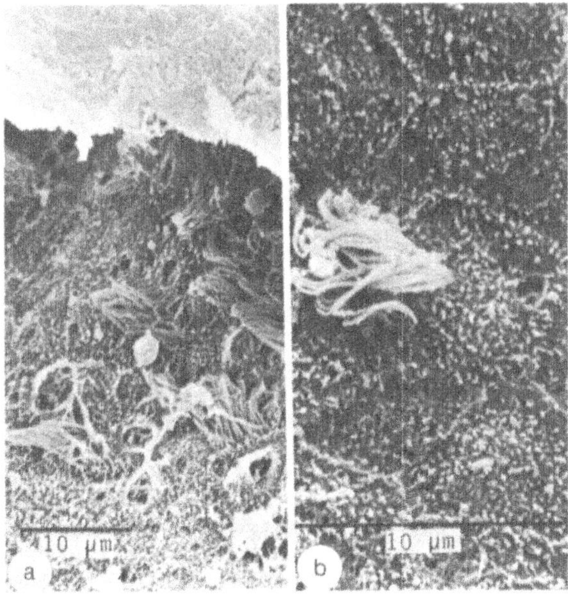

Fig. 3a, b. Glandular-cystic endometrial hyperplasia. **a** Glandular opening showing mucus-forming cells and ciliated cells. **b** Cystic gland showing expanded and flattened epithelial cells

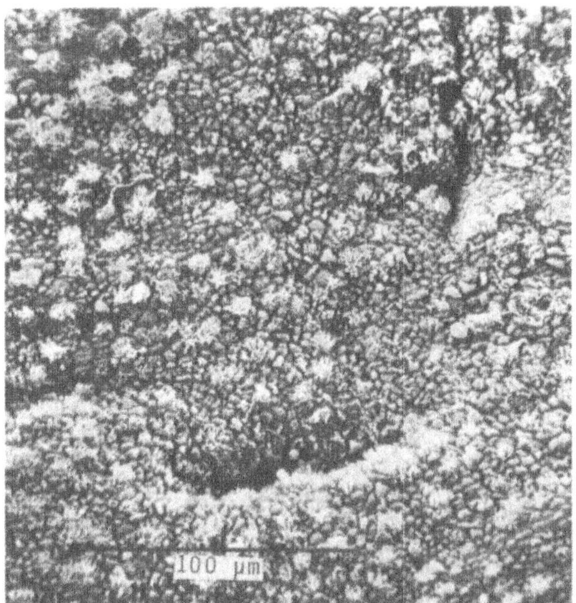

Fig. 4. Endometrium surface in a case of glandular-cystic hyperplasia

in irregular bunches and varied in shape and length, which is a typical finding in this disease (Ferenczy and Richart 1973a).

Ferenczy et al. (1972), Johannisson and Nilsson (1972), and Ferenczy and Richart (1973b) have all emphasized this characteristic distribution of ciliated cells, which they found concentrated around the openings of the ducts. We were

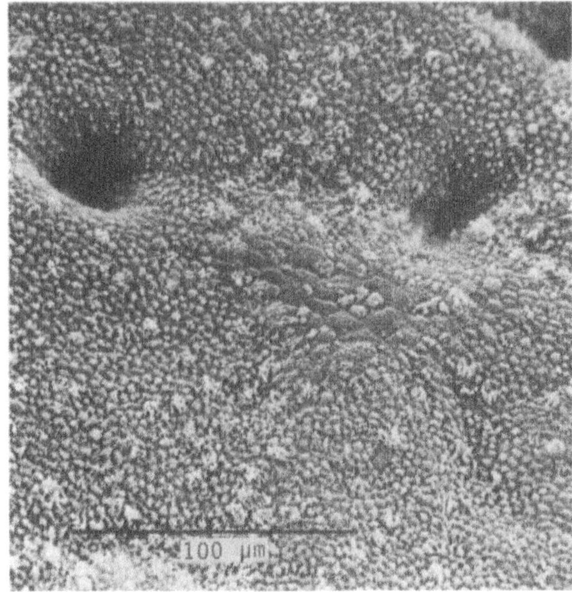

Fig. 5. Endometrium on 12th day of cycle. Note the irregular distribution of ciliated cells

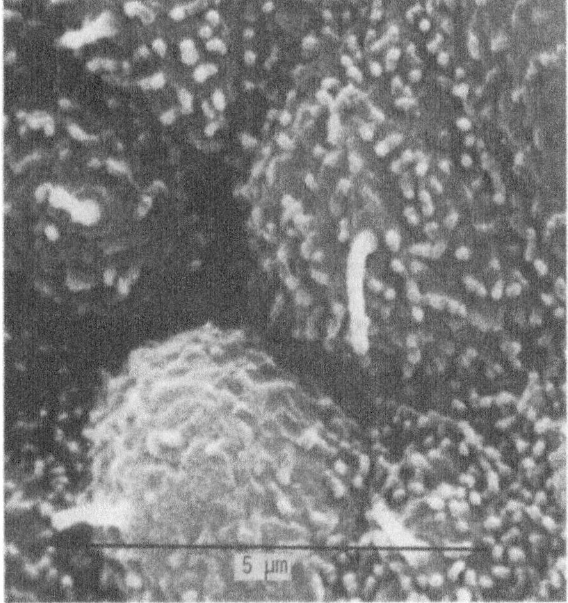

Fig. 6. Endometrium on 7th day of cycle. Note single cilia on cell surfaces

not able to corroborate this finding, however. Our samples showed a predominantly irregular distribution such as that seen on the endometrial surface from the 12th day of the cycle onwards (Fig. 5).

Being a changing tissue, the endometrium has a relatively high mitotic activity during its proliferation phase. Embryonic tissues as well distinguish themselves by numerous mitoses. Mitosis is stopped when a centriole wanders to the cell membrane and makes contact with the plasmalemm, whereupon a single cilium sprouts (Herbst 1975; Kober and Herbst 1975; Grabitz 1976; Herbst et al. 1976; Kober et al. 1976; Grabitz et al. 1977; Kober et al. 1977; Multier-Lajous and Herbst 1978; Schaller 1978). Ludwig and Metzger (1976a) illustrated cells with single cilia found on the 6th day of the cycle, saying they were ciliated cells whose significance they did not know. We too found cells with single cilia during the proliferation phase; since these cilia lacked central tubuli, they were stereocilia. We believe that the function of these single cilia is related to the stopping of mitosis (Fig. 6).

Böcker and Strecker (1975) and Böcker and Stegner (1975) have also detected single cilia on myoblasts from a leiomyosarcoma (or endometrial sarcoma). These and other findings of single cilia on tumor cells (Wilson and McWhorter 1963; Schuster 1964; Hashimoto et al. 1966; Anton-Lamprecht 1972; Kalderon and Tucci 1973; Yalciner and Friedell 1973) give rise to the question whether the appearance of single cilia might be an indication of tumor cell differentiation.

Summary

Changes in the surface relief of human endometrium under the influence of exogenous estrogen are shown in examples of adenomatous hyperplasia and glandular-cystic hyperplasia. Single cilia were detected on the endometrium surface during the proliferation phase; we believe these cilia induce cessation of mitosis.

References

Anton-Lamprecht I (1972) Intracelluläre Geißeln in Epidermiszellen menschlicher Haut. Arch Dermatol Forsch 243:199–208

Böcker W, Stegner H-E (1975) A light and electron microscopy study of endometrial sarcomas of the uterus. Virchows Arch [Pathol Anat] 368:141–156

Böcker W, Strecker H (1975) Electron microscopy of uterine leiomyosarcomas. Virchows Arch [Pathol Anat] 367:59–71

Ferenczy A (1976) Studies on the cytodynamics of human endometrial regeneration. I. Scanning electron microscopy. Am J Obstet Gynecol 124:64–74

Ferenczy A (1977) Surface ultrastructural response of the human uterine lining epithelium to hormonal environment. A scanning electron microscopic study. Acta Cytol (Baltimore) 21:566–572

Ferenczy A, Richart RM (1973 a) Scanning electron microscopy of hyperplastic and neoplastic endometria. Scanning Electron Microscopy 1973/ III, IITRI, Chicago, 613–620

Ferenczy A, Richart RM (1973 b) Scanning and transmission electron microscopy of the human endometrial surface epithelium. J Clin Endocrinol Metab 36:999–1008

Ferenczy A, Richart RM, Agate FJ, Purkerson ML, Dempsy EW (1972) Scanning electron microscopy of the human endometrial surface epithelium. Fertil Steril 23:515-521

Grabitz K (1976) Die Ultrastruktur des Epithels menschlicher Corpora ventriculi während der Ontogenese. Z Mikrosk Anat Forsch 90:577–609

Grabitz K, Herbst R, Kober H-J, Multier-Lajous A-M (1977) Zelldifferenzierung im menschlichen Magenepithel. Beitr elektronenmikrosk Direktabb Oberfl 10:525–530

Hafez ESE, Ludwig H, Metzger H (1975) Human endometrial fluid kinetics as observed by scanning electron microscopy. Am J Obstet Gynecol 122:929–939

Hashimoto K, Gross BG, Nelson RG, Lever WF (1966) Eccrine spiradenoma. Histochemical and electron microscopic studies. J Invest Dermatol 46:347–365

Herbst R (1975) Significance of scanning electron microscopy for embryological/teratological investigations. In: Neubert D, Merker HJ (eds) New approaches to the evaluation of abnormal embryonic development. Thieme, Stuttgart, pp 57–70

Herbst R, Kober H-J, Multier-Lajous A-M, Staubach J (1976) Die Bedeutung der Einzelcilien bei der Differenzierung. Beitr elektronenmikrosk Direktabb Oberfl 9:635–646

Johannisson E, Nilsson L (1972) Scanning electron microscopic study of the human endometrium. Fertil Steril 23:613–625

Kalderon AE, Tucci JR (1973) Ultrastructure of a human chorionic gonadotropin- and adrenocorticotropin-responsive functioning Sertoli-Leydig cell tumor (Type I). Lab Invest 29:81–89

Kober H-J, Herbst R (1975) Die lumenseitige Oberfläche des Rattenoesophagus während der Ontogenese. Z Mikrosk Anat Forsch 89:702–714

Kober H-J, Herbst R, Multier-Lajous A-M (1976) Oesophagus und Trachea der Ratte während der Ontogenese. Beitr elektronenmikrosk Direktabb Oberfl 9:615–624

Kober H-J, Schaller G, Herbst R, Grabitz K, Multier-Lajous A-M (1977) Zelldifferenzierung an der Lumenseite des menschlichen Oesophagus. Beitr elektronenmikrosk Direktabb Oberfl 10:509–516

Ludwig H, Metzger H (1976a) The re-epithelization of endometrium after menstrual desquamation. Arch Gynaekol 221:51–60

Ludwig H, Metzger H (1976b) The re-epithelization of the endometrium after menstrual desquamation. Scanning Electron Microscopy 1976/VI, IITRI, Chicago, 351–358

Ludwig H, Metzger H (1976c) The human female reproductive tract. A scanning electron microscopic atlas. Springer, Berlin Heidelberg New York

Ludwig H, Metzger H (1977) Ciliogenesis in the human fetal oviduct. Scanning Electron Microscopy 1977/ II, IITRI, Chicago, 361–366

Multier-Lajous A-M (1971) Gezielte Präparationsmethode für die Transmissions-Elektronenmikroskopie. Beitr Pathol 143:416–422

Multier-Lajous A-M, Herbst R (1978) Die Entwicklung der Nierenkörperchen beim Kaninchen. Beitr elektronenmikrosk Direktabb Oberfl 11:353–358

Nilsson O, Nygren KG (1972) Scanning electron microscopy of human endometrium. Ups J Med Sci 77:3–7

Schaller G (1978) Die Lumenseite des menschlichen Oesophagus während der Ontogenese. Z Mikrosk Anat Forsch 92:675–699

Schuster FL (1964) Ciliated fibroblasts from a human brain tumor. Anat Rec 150:417–422

Steger RW, Huang H, Kuppe G, Meites J, Hafez ESE, Ludwig H (1976) Effect of age on uterine ultrastructure. Scanning Electron Microscopy 1976/ VI, IITRI, Chicago, 359–366

Wilson RB, McWorther CA (1963) Isolated flagella in human skin. Lab Invest 12:242–249

Yalciner S, Friedell GH (1973) Cilia in the epithelium of the urinary bladder during experimental carcinogenesis. J Natl Cancer Inst 51:501–505

Effects of High Estrogen doses on the Endometrium

H. Egger and G. Kindermann

Abstract

The application of high dose estrogens (up to a total of 25 mg of ethinylestradiol in five days) resulted in marked endometrial edema with stromal lacerations and severe hyperemia, together with a retardation of the secretory transformation by 3 to 4 days. There were disturbances both of glandular mucus production and of reticular fiber structure in the second half of the cycle. The secretory transformation is arrested in the third to fourth day stage of a normal spontaneous cycle. The endometrium desynchronized by high dose ethinylestradiol therefore is not more a nidation-endometrium. Similar endometrial pattern (like those after high dose ethinylestradiol) were seen in rhesus monkeys given postcoital doses of estrogens and infertile woman with atypical secretory endometria. These morphological changes seem to be the basis of nidation-inhibition by the "morning after pill."

Empirically, we know that extremely high doses of estrogen taken postcoitally are an effective, emergency contraceptive. A dose of 5 mg of ethinylestradiol has proven effective when taken on 5 consecutive days starting 24 h at the latest after cohabitation. The mechanism of action of these high dose estrogens however is unknown. Given before ovulation they, of course work blocking. We were interested in determining the effect they have on postovulatory endometrium, in morphological terms.

The well-known effects of low, physiologic amounts of estrogen on the endometrium were also seen with the "morning after pill", i.e., proliferation of glands and stroma, hyperemia, increased vascularization, and edema formation. All of these typical estrogen effects were markedly increased by extremely high doses of ethinylestradiol. The formation of edema, in particular, led to stromal laceration and lacunae of extravasation in the upper endometrium to the point of hemorrhage.

Above and beyond this, high doses of ethinylestradiol (e.g., 3–5 mg/24 h on 5 consecutive days) effected very specific changes in the secretion pattern. The secretory transformation of the postovulatory endometrium was arrested at the stage of generalized, retronuclear vacuoles characteristic in the normal cycle (only for a short time) for the 4th day postovulation. These generalized, retronuclear vacuoles covered the entire edematous and blown up edometrium, and continued under high doses of estrogens into the 8th, 9th, and 10th days post ovulation.

During these conditions, the endometrium obviously cannot serve for the nidation of the embryo. This endometrial pattern is clearly very different from that normally seen on the 6th to 8th day postovulation. This is particularly true hormonal comparing two endometria samples of the 7th day postovulation, one without treatment and the other after the "morning after pill" (after 5 days higher dose estrogene).

Parallel to these morphological changes caused by extremely high doses of estrogens a decreased pregnanediol secretion is observed during the first and second secre-

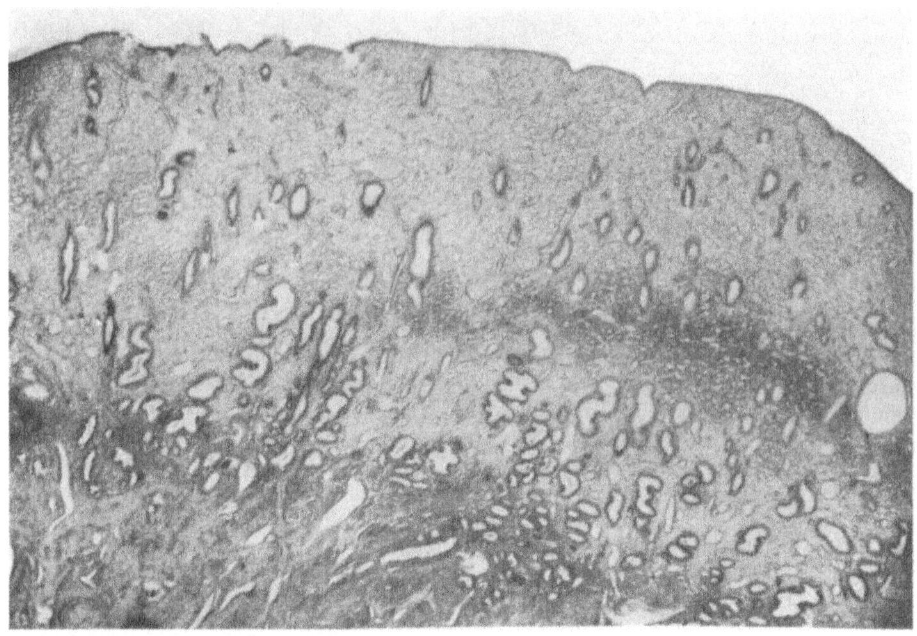

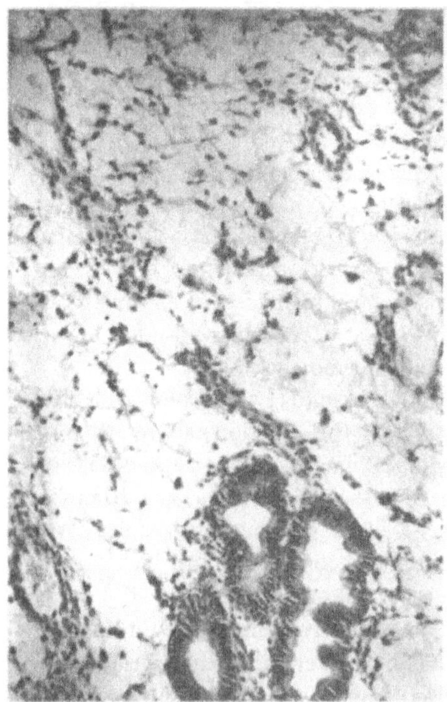

Fig. 1. Endometrium of the 8th day post ovulation after 5 x 5 mg E_2 given per mouth

tion phases. This effect is also know as "luteolysis", but may be explained otherwise. Once nidation has occured ethinylestradiol in the same amounts do not have any adverse effect on embryonal development in spite of the obvious histologic changes it causes in the endometrium.

Summary

Administration of high doses of estrogen (up to 25 mg of ethinylestradiol over 5 days) resulted in marked endometrial edema with stromal lacerations and severe hyperemia. Secretory transformation was retarded by 3 to 4 days. Disturbances in both glandular mucus production and reticular fiber structure were noted in the second half of the cycle. Endometrium thus desynchronized cannot give nidation place. Patterns similar to those seen in the human endometrium after high doses of ethynylestradiol were also seen in rhesus monkeys given postcoital estrogens, and in infertile women with atypical endometrial secretion. These morphological changes seem to be the mechanism of action inhibiting nidation by the "morning after pill."

Retrospective Histologic Examination of Endometrium in Curetting and Hysterectomy Specimens Following Estrogen Therapy

U. Bonk

Abstract

Elevated levels of estrogen with low progesterone levels may develop for prolonged periods during estrogen therapy for menopausal complaints, for gonadal dysgenesis, or pituitary deficiency, for the treatment of breast cancer or as oral contraceptives. We present from a retrospective study of women receiving such therapy the histologic changes that occurred in their endometrium (curettings) or later in their uteri, extirpated for various reasons.

The histologic changes that we observed over a period of more than 3 years are recorded in an alphabetically arranged card catalog of the patients; that system allowed us to record information about previous therapy prescribed by other clinicians in a community of 100,000 persons. The various forms of endometrial hyperplasia we saw in the curettings are of great interest, since such changes are not uniformly evaluated or classified in the literature, and they require close cooperation with the gynecologist. We placed special importance on trying to differentiate histologically between reversible atypical endometrial hyperplasia and true endometrial carcinoma.

It is worth noting that a pronounced endometrial hyperplasia in curettings rarely was correlated with similar prognostically uncertain changes in the extirpated uterus. We suggest that a histologic grading system of definite markers or criteria be devised, so that the changes can be better classified, compared, and treated.

Terminology

A number of different classifications have been suggested for the endometrial hyperplasias. Table 1 lists a selection of the best known of these. Following the early work of Campbell and Barter (1961), and recent publications by Welch and Scully (1977), we have chosen to classify hyperplasias as cystic and atypical, and to further differentiate them by degree (low grade I, medium grade II, high grade III), depending on their *structure* and *cytology*. This gives us sound criteria by which to evaluate histologic material. We do not use the term adenomatous hyperplasia since not enough is known about its precancerous potential.

Cystic hyperplasia

Glands of varying size, some of which are cystically expanded, with thick epithelium of one to two layers which is sharply bordered at the lumen. Cell-rich stroma with "nude nuclei." Expanded sinusoids with thrombosis and small infarcts.

Table 1. Endometrial hyperplasia – classifications

Campbell and Barter 1961	Gusberg and Kaplan 1963	Beutler et al. 1963	Gore and Hertig 1966	Vellios 1972	WHO 1975	Own
Benign hyperplasia	Slight adenomatous hyperplasia	Cystic proliferation	Cystic hyperplasia	Cystic hyperplasia	Cystic hyperplasia	Cystic hyperplasia
Atypical hyperplasia Type 1	Moderate adenomatous hyperplasia	Glandular hyperplasia	Adenomatous hyperplasia	Adenomatous hyperplasia	Adenomatous hyperplasia	Low grade atypical hyperplasia, I
Atypical hyperplasia Type II			Anaplasia	Atypical hyperplasia		Medium grade atypical hyperplasia, II
Atypical hyperplasia Type III	Serious adenomatous hyperplasia	Glandular hyperplasia with atypical epithelial proliferation	Carcinoma in situ	Carcinoma in situ	Atypical hyperplasia	High grade atypical hyperplasia, III

Atypical hyperplasia I, low grade (Fig. 1, 2)

Structurally, enlarged glands containing small polyps which are histologically detectable at low magnification. Glands may be crowded ("adenomatous hyperplasia"). Cytologically, slightly atypical cell shapes usually, but not always, accompanied by atypisms in structure. Alterations in the nuclei such as rounding off, size variation, and high susceptibility to dye are also characteristic, as is a layering of the glandular cells.

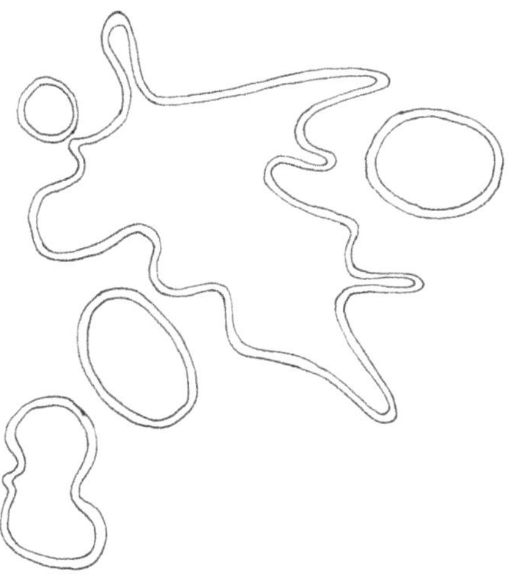

Fig. 1. Cystic and atypical (architectural) hyperplasia (I grade architecture)

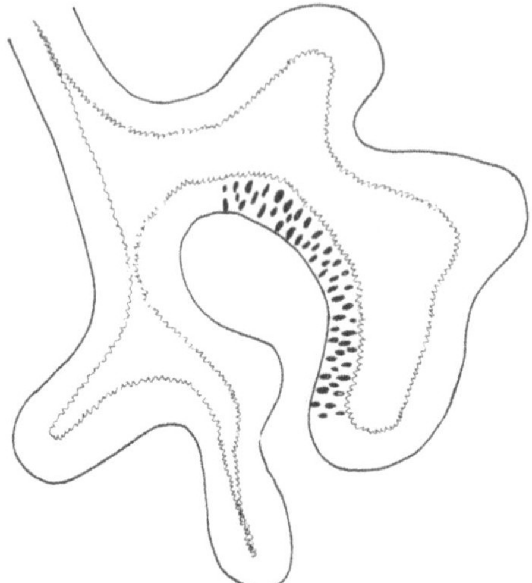

Fig. 2. Atypical (architectural) hyperplasia with slight cellular atypia (I grade architecture, I grade cytology)

Atypical hyperplasia II, medium grade (Fig. 3)
Structurally and cytologically intermediate between grades I and III.

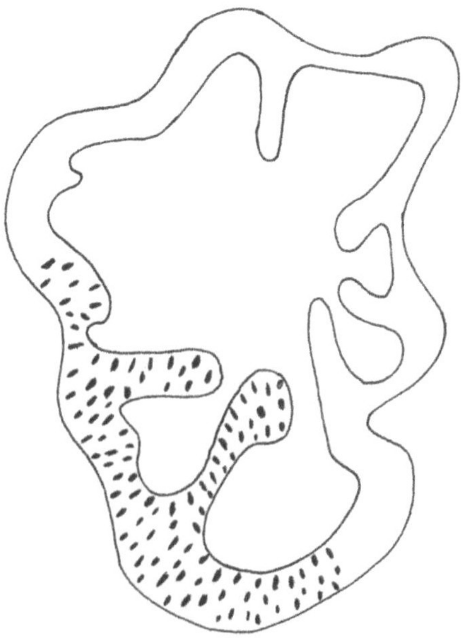

Fig. 3. Moderate atypical hyperplasia with architectural and cellular atypia (II grade architecture, II grade cytology)

Atypical hyperplasia III, high grade (Fig. 4, 5)

Structurally, we begin to see intraglandular proliferation, which may lead to blockage of the interspaces. The glands may lie back to back. Cytologically, we see considerable nuclear alteration including variation in shape and size, loss of polarity, prominent nucleoli, strong dye susceptibility or granular chromatin, and a high degree of glandular cell layering. The cytoplasm edge may be broad and eosinophilic. Cellular atypisms may be present even though structural abnormalities are not (Hertig et al. 1949).

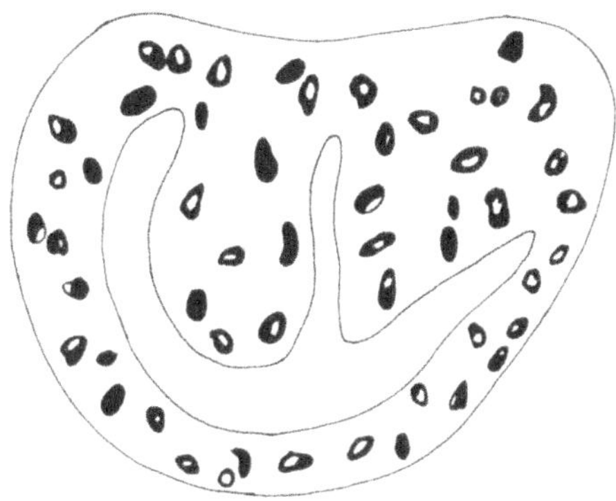

Fig. 4. Severe atypical hyperplasia with intraglandular cellular tuft (III grade)

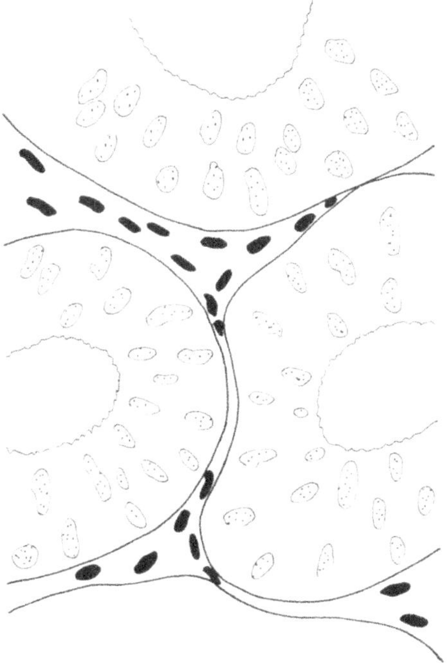

Fig. 5. Severe atypical hyperplasia. Cells with abundant eosinophilic cytoplasm and with nuclei containing granular chromatin (III grade)

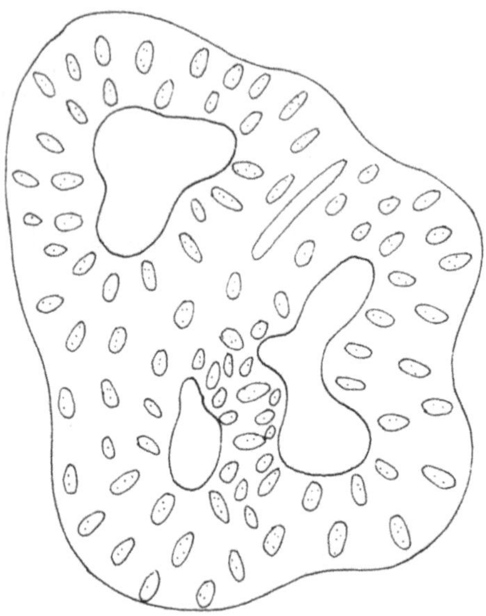

Fig. 6. Carcinoma in situ Intraglandular bridges and severe atypia are present (III grade cytology)

"Carcinoma in situ" (Fig. 6)
Intraglandular bridging, high grade cellular atypism, and focal attack on not more than five or six glands. Nuclear membrane is irregular and contains enlarged nucleoli with uneven edges or spurs. To date we know nothing about the malignant potential of atypisms in structure and cytology. We would recommend the method of Welch and Scully (1977) to document and evaluate findings.

Method and Results

In 6000 curettings we found 333 cases of endometrial hyperplasia in women who had had estrogen treatment for menopausal complaints. This sample was highly heterogenous, having come from private practices and stationary hospital wards. The patients' ages were between 40 and 65 years. Histologic analysis of the material revealed 244 cases of cystic hyperplasia (CH), 45 of atypical hyperplasia, grade I (AH I), 32 of atypical hyperplasia, grade II (AH II), and 12 of atypical hyperplasia, grade III (AH III). In eight cases of endometrial carcinoma found there was a clinical history of estrogen treatment (ET).

Uterus extirpation was performed in the above cases a year at the latest following curettage and diagnosis. In 57 of the 244 cases of cystic hyperplasia we removed the uterus immediately after diagnosis; 34 showed a normal endometrium and 23 cystic hyperplasia. A comparison of the other curet and uterus diagnoses is given in Table 2.

This data indicates that in none of the cases studied did endometrial hyperplasia develop into carcinoma between the time of diagnosis and extirpation. Quite the contrary, many of the uterus samples from cases of endometrial hyperplasia showed no, or only focal, remains of the disease. There was often a lower grade of atypical hyperplasia present, and frequently only cystic hyperplasia or even "normal" endometrium.

Table 2. Number of cases and results by examination of endometrium in curettings and hysterectomy specimens

Curettage		Uterus		
CH	244	34 ET	23 CH	187 ?
AH I	45	23 ET	10 CH	12 AH I
AH II	32	11 OB	5 CH	16 AH II
AH III	12	4 CH	8 AH III	
CA	8	8 CA	(3 IN SITU)	

The endometrial carcinomas we found were highly differentiated adenocarcinomas which had penetrated a maximum of 2 mm into the myometrium. The uterus was extirpated in these cases immediately after diagnosis. In three cases we found only the remains of focal, well differentiated adenocarcinoma (carcinoma in situ).

These cases were chosen according to physicians' reports that the women had been taking estrogens for menopausal complaints, prescribed either by them or other doctors. Here, of course, we were faced with the problem of erratic estrogen pretreatment of the type given in general practice. In order to obtain more information on these women, we asked their gynecologists to answer a number of questions on the treatment they had been giving and its effects on the endometrium, indicating that we as pathologists knew nothing about frequency of estrogen use and its relation to the many other patients who showed no functional or other complications after estrogen therapy. Each gynecologist kindly provided us with a list of an average of 500 patients who had taken estrogens during menopause for long periods of time (i.e., 6 to 36 months). In the main, these were estrogen-androgen combinations, conjugated estrogens, estradiol, estriol, and others. In only 1% to 5% of women treated with these substances were *conspicuous* changes observed.

The symptoms these women suffered from were nausea, headaches, portio alterations due to gestagen component, growth of facial hair, voice changes (rare), mastodynia, and hemorrhaging. Serious disorders following estrogen therapy were seen only *extremely rarely*. The gynecologists also remarked that they could not always say precisely how often their patients had taken estrogens or when they had stopped, since both general practitioners and internists had often hindered the conscientious use of the drugs "for emotional reasons." They suspected, however, that long-term estrogen treatment with injections of, say, estrogen-androgen combinations (Gynodian Depot) might well lead to atypical endometrial hyperplasia and perhaps even to endometrial carcinoma.

Discussion

In contrast to Gusberg and Kaplan (1963), and like Buehl et al. (1964), we found in the extirpated uteri no focal endometrial carcinoma due to atypical endometrial hyperplasia. The conspicuous regression of hyperplasia that we saw, may have been due to ongoing gestagen therapy, cessation of estrogens and the trauma of curettage.

Having had no control group, it was not possible for us to determine whether estrogens had been an etiologic factor in the development of endometrial hyperplasia and carcinoma. The fact that in none of our cases did we see a progression of atypical endometrial hyperplasia after curettage and gestagen therapy, however, and that the suspected endometrial carcinomas were well differentiated and not highly aggressive, speaks on the whole for a good prognosis of endometrial hyperplasia after estrogen therapy in menopause. This is corroborated by many authors (e.g., Tavassoli and Kraus 1978).

What the practicing gynecologists told us about having only very rarely observed serious disorders following estrogen treatment in menopause, lends support to the most recent publications on the subject in German-speaking countries. These reports indicate that the way the drugs are prescribed here – cyclic substitution if at all possible, a gestagen component, and lowest effective doses for only as long as needed – seems to give no reason for objection (Lauritzen 1978; Völker et al. 1978).

Since we know that spontaneous, atypical hyperplasias are often reversible by gestagen therapy (Wentz 1966; Steiner et al. 1965), it is probable that developing endometrial carcinoma has a long prodromal phase which depends on exogenous or indogenous promotors. The short observation time in our cases does not allow us to prognosticate, yet it does throw a light, we think, on the situation in daily practice, outside the controlled studies in the universities. The drugs that have aroused suspicion in the United States are primarily conjugated estrogens; and this is the type the gynecologists we polled had most often prescribed. On the other hand, almost all of these physicians thought that injections of estrogen in depot form in combination with androgens and other substances, of the type preferred by general practitioners, were perhaps more of a risk, since they are repeatedly brought into connection with individual cases of endometrial carcinoma and the associated anamnesis.

Theoretically, the suspicion that estrogen-androgen combinations might lead to endometrial carcinoma, is lent some support by the work of MacDonald et al. (1971), who say that even synthetic steroids may be transformed to estron in extraglandular fat tissue. Depending on the concentration of hormone receptors in the endometrium, plasma steroid level, and probably other factors, estron may affect the target organ of endometrium and lead, depending on its functional state, to diffuse or limited (polyp) proliferation (Schindler 1977). In the United States, the method used to statistically evaluate retrospective studies on the estrogen-endometrial carcinoma problem, has recently been subjected to much criticism (Feinstein and Horwitz 1979; Hutchinson 1979).

We believe that the recommendations of the Drug Commission of the German Association of Physicians (1978) should be followed, in order to keep the risk of malignant tumors as small as possible – at least for as long as early detection is the best we can do, and prevention remains utopian.

I should like to express my deep thanks to my colleagues for their cooperation and generous provision of data, for answering our questions frankly, and for all the other information they so kindly provided.

References

Bekanntmachung der Arzneimittelkommission der Deutschen Ärzteschaft (1978). Weibliche Sexualhormone und Endometriumkarzinom. Dtsch Ärzteblatt 75:1306

Beutler HK, Dockerty MB, Randall LM (1963) Precancerous lesions of the endometrium. Am J Obstet Gynecol 86:433–443

Buehl IA, Vellios F, Carter JE (1964) Carcinoma in situ of the endometrium. Am J Clin Pathol 42:544–601

Campbell PE, Barter RA (1961) The significance of atypical endometrial hyperplasia. J Obstet Gynaecol Br Commonw 68:668–672

Feinstein AR, Horwitz RI (1979) Correspondence N Engl J Med 300:497

Gore H, Hertig AT (1966) Carcinoma in situ of the endometrium. Am J Obstet Gynecol 94:134–155

Gusberg SB, Kaplan A (1963) Precursors of corpus cancer. IV.Adenomatous hyperplasie as stage 0 carcinoma of the endometrium. Am J Obstet Gynecol 87:662–676

Hertig AT, Sommers SC, Bengloff H (1949) Genesis of endometrial carcinoma. Part III. Carcinoma in situ. Cancer 2:964–971

Hutchinson GB (1979) Correspondence. N Engl J Med 300:497

Lauritzen C (1978) Östrogene und Endometriumkarzinom. Fortschr Med 96:2293–2294

MacDonald PC, Grodin JM, Siiteri PK (1971) Dynamics of androgen and estrogen secretion. In: Baird DTI Strong (eds) Control of gonadal steroid secretion. University Press, Edinburgh, p 76

Schindler AE (1977) Endometriumkarzinom und extraglanduläre Östrogenbiosynthese. Geburtshilfe Frauenheilkd 37:242–251

Steiner GJ, Kistner RW, Craig JM (1965) Histological effects of progestins on hyperplasia and carcinoma in-situ of the endometrium – further observations. Metabolism 14:356

Tavassoli F, Kraus F (1978) Endometrial lesions in uteri resected for atypical endometrial hyperplasia. Am J Clin Pathol 70:770–779

Vellios F (1972) Endometrial hyperplasias, precursors of endometrial carcinoma. In: Sommers SC (ed) Pathology annual. Appleton-Century-Croffs, New York, pp 201–229

Völker W, Kannengießer U, Majewski A, Vasterling HW, (1978) Östrogentherapie und Endometriumkarzinom. Geburtshilfe Frauenheilkd 38:735–743

Welch WR, Scully RE (1977) Precancerous lesions of the endometrium. Hum Pathol 8:503–512

Wentz WB (1966) Treatment of persistent endometrial hyperplasia with progestins. Am J Obstet Gynecol 96:999–1004

WHO-Classification (1975) Histological typing of female genital tract tumours. Poulsen HE, Taylor CW. Geneva

Cytologic Effects of Estrogen Tests on Postmenopausal Vaginal and Cervical Epithelia

Y.P. Clocuh

Abstract

The action of the follicle hormone on squamous epithelium and on glandular epithelium depends on its pharmacologic composition and on its dosage, and can best be measured on the atrophic epithelium of the female genitalia.
Four hundred and ninety-two postmenopausal women were treated for inflammatory or equivocal cell-pictures with low doses of estrogen preparations. The action of estrogen on the squamous epithelium was characterized by the high build up of the epithelium and by the appearance of lipid granules in the cytoplasm of superficial cells.

Glandular epithelial cells from the endocervix were twice as common in the repeated smears and revealed large vesicular nuclei. In no instance did we detect endometrial cells in the smears after these low doses of estrogen.

Besides this evidence of regular cell proliferation there were also benign cellular alterations:

1. Nuclear enlargements, seen in 4% of the cases. These did not indicate dysplasia, however.
2. In 4% of the cases the cell picture was that typical of parakeratosis occurring after estrogen clearing in old women, but which may be confused with premalignant or malignant epithelial changes.
3. In some cases squamous epithelial cells were seen with single or double nuclei surrounded by large spaces, similar to those occurring in folic acid deficiency or in condylomata.

All of these benign changes disappeared after a few weeks, and at latest 3 months.

The effects of different estrogen derivates on squamous and glandular epithelia of the female reproductive tract depend on pharmacologic composition as well as on applied doses (Carlborg 1969; Dosne de Pasqualini 1950; Gitsch et al. 1960; Haskins et al. 1968; Hustin and Van der Aynde 1977; Lauritzen 1975). These effects can be studied most simply with atrophic vaginal epithelium. Small doses of estradiol derivates are sufficient to induce proliferation of vaginal epithelium, cervical glandular cells, or endometrial cells in postmenopausal women. Even bleeding can sometimes be observed. Pure colpotropic estriol preparations can be applied in higher doses according to Puck et al. (1957) without producing side effects.

From 1971 to 1973 estrogen-containing preparations were given to 492 postmenopausal women in order to clarify atrophic inflammatory catologic findings. A single dose of 2.5 mg diethylstilbestrol dipropionate (Cyren B forte) was administered to 95% of the patients. Five percent of the patients received oral or local medications of

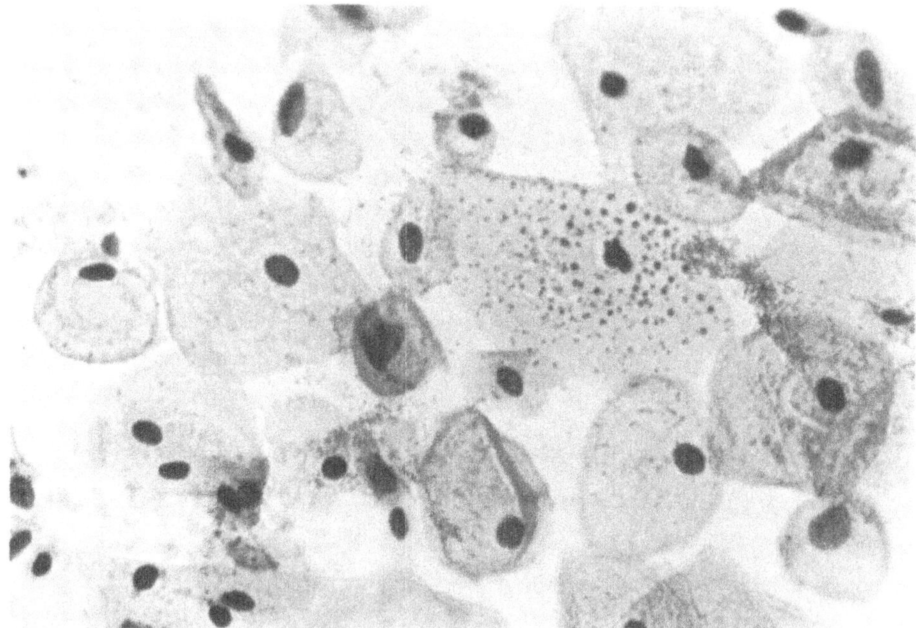

Fig. 1. Highly differentiated squamous epithelia with lipidgranula in the cytoplasm. x 456.5

other estrogen-containing pharmaca. Cytologic smears taken after estrogen treatment showed in all women highly differentiated squamous epithelia with lipidgranula in the cytoplasm (Fig. 1). Lipidgranula are looked upon as indicating a good degree of estrogen reaction.

Endocervical glandular epithelia were found twice as frequently. These epithelia were embedded in mucus and exhibited large vesicular nuclei with prominent nucleoli (Fig. 2). It should be pointed out, that under administration of these small doses of estrogen no endometrial cells were found in cytologic smears. This is an advantage in cytologic assessment of smears.

Besides of these normally proliferating epithelia benign cellular changes were observed. In 4% of the cases squamous epithelial cells of upper and intermediate strata were present, showing marked nuclear enlargement (Fig. 3). These changes, which closely resemble a slight dysplasia, disappeared within 6 weeks to 3 months in 80% of the cases. Moreover, single superficial and intermediate squamous cells were found exhibiting large perinuclear halos, sometimes showing binucleation (Fig. 4).

Similar changes were found in animal experiments by Dosne de Pasqualini et al. (1950) to be due to estrogen. But they can also occur in cases of folic acid deficiency or condylomatous alterations.

Another 4% of the cases showed parakeratosis after an estrogen test (Clocuh, to be published). Smears showed a clean background (Fig. 5). Well-differentiated squamous epithelial cells and numerous metaplastic cells showing different degrees of maturation with excentrically located nuclei were found. Dispersed across the slides we

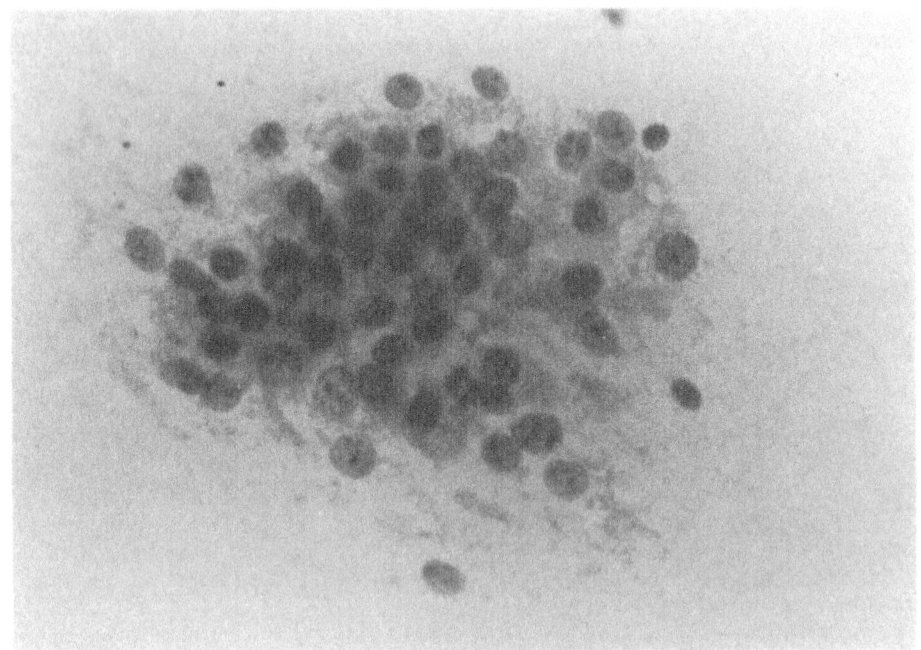

Fig. 2. Endocervical glandular epithelia with vesicular nuclei and prominent nucleoli. x 447.5

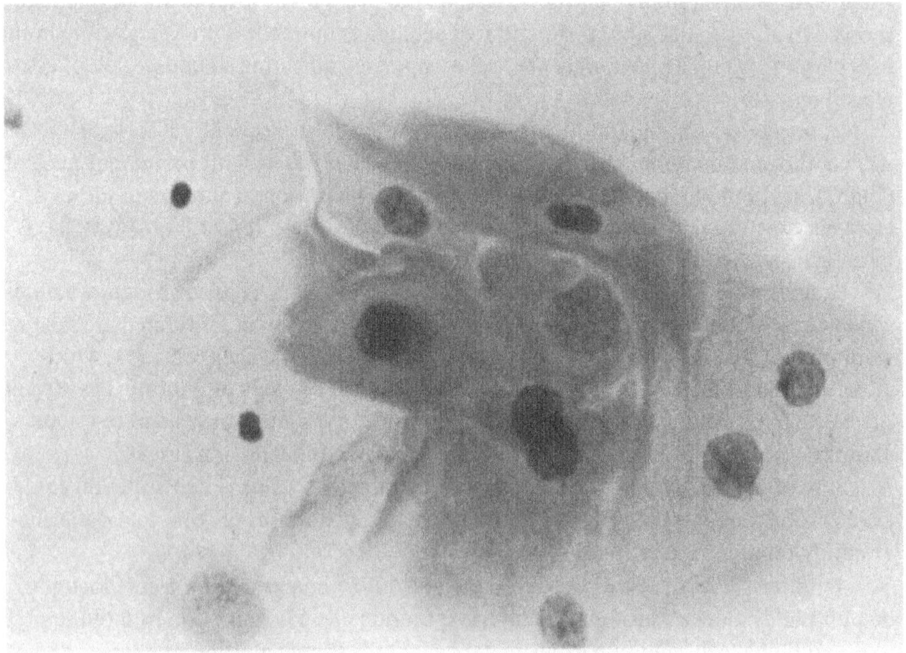

Fig. 3. Squamous epithelia showing marked nuclear enlargement. x 447.5

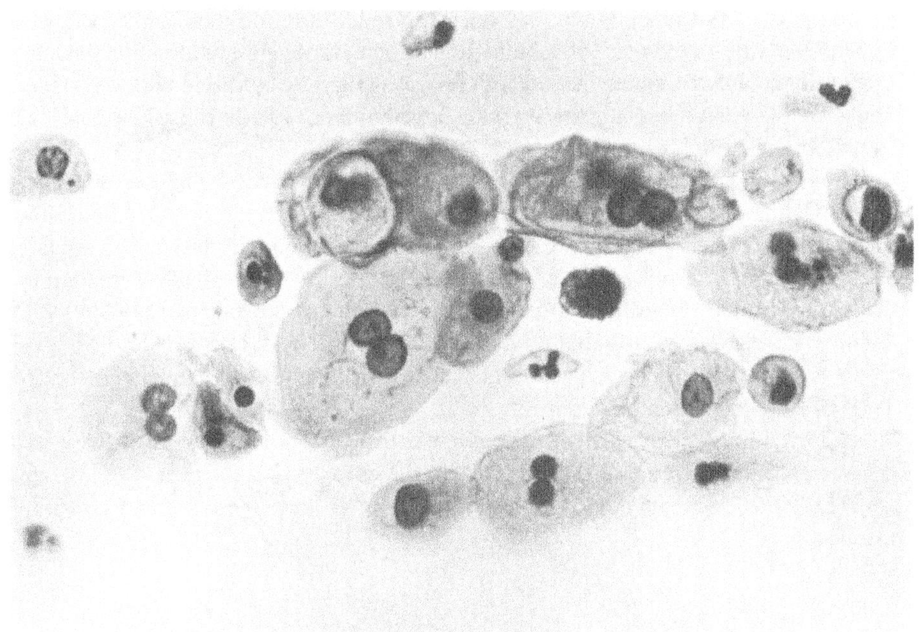

Fig. 4. Binucleated squamous epithelia. x 456.5

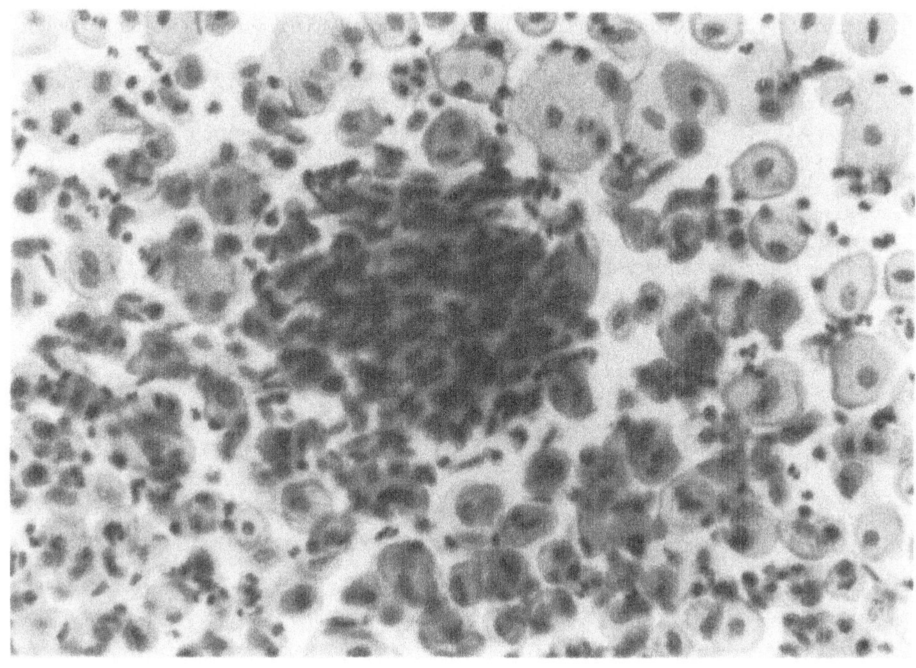

Fig. 5. Numerous metaplastic cells and parakeratotic cells. x 267

found crowded and singly lying small cells with small rims of cytoplasm. Nuclei were hyperchromatic, round-oval, or spindle-shaped, sometimes closely resembling nuclei of immature metaplastic squamous cells. These cells might be confused with cells from immature carcinoma in situ. However, the large number of these cells should be a clue to correct diagnosis.

Generally, we recommend in these cases a control smear to be taken after repeated estrogen treatment. Atypical cells hidden between parakeratotic cells would show up after this procedure, because benign cells of this type do not persist after repeated estrogen test. Our results indicate, that an estrogen test is well-suited as a method in diagnostic cytology in inflammatory, degenerative changes of vaginal squamous and endocervical glandular epithelium. Minimal doses of different estrogen-containing medicaments can be used successfully. Estriol medicaments can be recommended because of their colpotropic effects.

Benign cellular changes might be observed even with minimal doses of these medicaments, sometimes representing difficult problems in cytodiagnosis.

Therefore, a repeated cytologic examination after hormone treatment is recommended.

References

Carlborg L (1969) Action of diethylstilboestrol on mouse vaginal sialic acids. Acta Endocrinol (KBH) 62:657

Clocuh YP (to be published) Über das Auftreten von Parakeratosezellen in Zervikalabstrichen nach hormonaler Aufhellung und seine Bedeutung. Fortschritte Med

Dosne de Pasqualini CD, Terzano G, Buz GE (1950) Estrogenos y extendido vaginal. Endocrinologia Buenos Aires 1:77

Gitsch E, Müller-Hartburg W, Homolar W (1960) Potenzierung der lokalen Wirkung von Östriol. Geburtshilfe Frauenheilkd 20:1052

Haskins AL, Moszkowski EF, Whitelock VP (1968) The estrogenic potential of estriol. Am J Obstet Gynecol 102:665

Hustin J, Van der Aynde JP (1977) Cytologic evaluation of the effect of various estrogens given in postmenopause. Acta Cytol (Baltimore) 21:225

Lauritzen C (1975) Grundlagen der hormonalen Therapie klimakterischer Beschwerden. Dtsch Ärzteblatt 4:205

Puck A, Korte W, Hübner KA (1957) Die Wirkung des Oestriol auf Corpus uteri, Cervix uteri und Vagina der Frau. Dtsch Med Wochenschr 82:1864

Clinical Effects of Estrogens

H. Wittlinger

Abstract

The indications for estrogen treatment in females of different ages are presented and discussed. In particular, we approach critically the use of estrogen to reduce the height of potentially very tall young girls and the administration of estrogen-containing contraceptives to girls with irregular menstrual cycles. A large part of the paper deals with estrogen treatment in the pre- and postmenopause. The side effects are listed, and contraindications and diagnostic measures during the treatment are presented.

Orally or parenterally administered estrogen-containing substances affect various organs in the human female. The main target organ is the uterus, where estrogens affect the myometrium, endometrium, and endocervix. Other target organs are vaginal epithelium, oviductal mucosa, skin and subcutaneous skin layers, breast, and finally, by way of hormonal metabolism, the pituitary and diencephalon (Fig. 1).

The indications for treatment with estrogen containing substances vary considerably, and usually depend on the age of the patient. The youngest patients for whom clinicians sometimes prescribe estrogen therapy are girls of 10 – 11 years who show signs of growing too tall. By giving high dosages of estrogens over a period of several months, one can reduce their final height by about 5 cm. This treatment is usually prescribed only if the girl's height can be expected to be over 190 cm. The disadvantages of this treatment are premature sexual maturation, the necessity of frequent gynecologic examinations, and finally a higher risk of cancer, which cannot be completely ruled out.

A further common indication for estrogen treatment are irregular, usually anovulatory, cycles in 14- to 18-year-old girls. Here it should be stressed that the contraceptives often prescribed for this condition may lead to regular menstrual cycles and thus be apparently successful; when they are discontinued, however, they usually leave an anovulation with amenorrhea or aligomenorrhea, which is very difficult to treat.

Estrogens are most widely used in sexually mature women for purposes of contraception. There is no longer any disagreement that women who take estrogen-containing contraceptives, particularly if they smoke cigarettes, run a higher risk of thrombosis. The maximum single dose is considered to be 0.05 mg.

Another large area of application of estrogen-containing drugs is the treatment of menopausal and postmenopausal complaints. In the case of women of about 45 years of age and older who are in premenopause, i.e., before menstrual bleeding has ceased, treatment has the following aims:

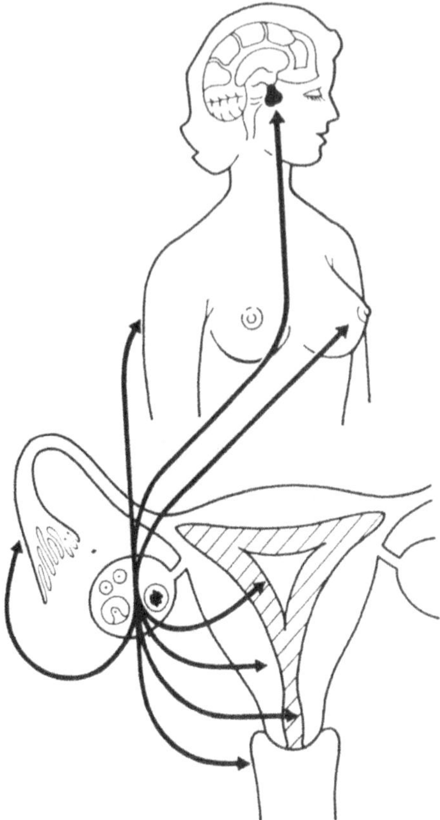

Fig. 1

1. Regulation of the cycle.
2. Contraception
3. Relief of vegetative disorders.

In women who are in postmenopause, i.e., after at least one year of amenorrhea, the prime goals of estrogen treatment are:

1. Relief of vegetative disorders
2. Treatment of trophic disorders in the urogenital tract
3. Prophylaxis of degenerative alterations of the skin, arteries, and skeletal system.

To illustrate how the biologic aging process has changed in women, and how women's place in society has changed over the centuries, completely altering the demands made on medical science, I have juxtaposed two pictures – Dürer's portrait of his mother (Fig. 2) and a typical "young grandmother" of today (Fig. 3).

We have available for treatment in premenopause a number of different drugs with varying proportions of estrogens and gestagens (Table 1). Using these drugs, we can regulate the usually unstable and often anovulatory cycles that women have at this period of their life. These drugs are also reliable contraceptives and very useful in eliminating any vegetative disorders these women may suffer from.

To treat vegetative and trophic disorders during the climacteric and postmenopause, we can choose from among the many different estrogen-containing drugs. These

Fig. 2

have varying effects on the psyche, vegetative nervous system, vaginal epithelium, endometrium, and the epithelium of the urinary tract (Table 2). Methylestradiol, for example, has both a strong central and peripheral effect. Its side-effects include nausea, a feeling of tension in the breasts, edema, and often breakthrough bleeding. Estradiol valerianate has favorable effects on the psyche and the neurovegetative system and

Fig. 3

Table 1. Effects and dosages of frequently used oral estrogens (modified after Lauritzen)

Substance	Trade name	Effects	Threshold dose over 2 weeks (mg)
Ethynylestradiol	Estrovis	Strong central and peripheral effects, nausea, tenderness of the breasts, edema, frequent bleeding cholesterol-reducing	0,2
Estrodiol valerate	Progynova Oestrogynal	Well tolerated, favorable neurovegetative and psychic effects, centrally weaker, minor side effects, cholesterol-reducing	6 – 10
Conjugated estrogens	Presomen Transannon Oestro-Feminal	Similar to estrodiol valerate, good psychotropic effect, cholesterol reducing, possible increases in triglyceride and phospholipids	
Estriol	Ovestin Synapause Hopmomed Crynaesan	Weak central and endometropic effect, favorable influence on the vegetative system and on the vulva, vagina, urethra, and bladder	20

has few side-effects. Conjugated estrogens similarly have a good psychotrophic effect, and have been shown to lower cholesterine levels and sometimes to increase those of tryclicerides and phospholipids. Drugs that contain estriol evince weak central and endometriotrophic effects, and favorably influence the vegetative system and the epithelium of vulva, vagina, urethra, and bladder.

When administering estrogens we have to take a number of significant side-effects into account:

1. Tendency to edema formation, weight increase, tension in the breasts, mastodynia.
2. Vasomotor symptoms, increase in blood pressure, headaches, migranes, dizziness, impaired vision.
3. Changes in cholesterine and lipid metabolism, increase in phosopholipid and trycliceride levels, lowering of cholesterine level.
4. Influence on clotting factors, increased risk of thromboembolism.
5. Withdrawal bleeding in 1 – 5% of cases, abrasio necessary to prevent malignoma.

Contraindications to estrogen treatment are the following:

1. Absolute: History of thrombosis, serious hepatopathy or latent hepatitis, pregnancy hepatosis, Rotor-Dubin-Jonson syndrome, serious hypertonia, brain vessel disorders, breast cancer, endometrial cancer.

Table 2. Compound preparations used in treatment during the postmenopause

Trade name (manufacturer)	Composition	Dose (mg)
Nuriphasic (Nourypharma)	Ethynylestradiol Lynestrenol	0.05 2.5
Duoluton (Schering)	Ethynylestradiol Norgestrel	0.05 0.5
Neogestakliman (Asche)	Ethynylestradiol Norethisterone acetate	0.05 2.0
Trisequens (Novo)	Estradiol Estriol Norethisterone acetate	2.0 1.0 1.0
Presomen comp. (Kali-Chemie)	Conjugated estrogens Medrogestone	1.25 5.0
Cyclo-Progynova (Schering)	Estradiol valerate Norgestrel	2.0 0.5

2. Relative: Diabetes mellitus, migrane, epilepsy, cardial and nephrogenic edema, myoma, endometriosis, mastopathy, tuberculosis.

Considering the possible side effects and the above contraindications, it is wise before starting any type of estrogen treatment to make a very careful check of the patient's case history, age, and indications. During treatment it is absolutely necessary to perform a complete general and gynecologic anamnesis, and a gynecologic examination every 6 months including vaginal smear for cytologic tests; tabs should also be kept on the patient's weight and blood pressure.

Finally, estrogens should not be the primary or sole method of treating vegetative and trophic disturbances that occur during and postmenopause. Equal weight should be given to dietary suggestions, physical measures, and especially psychologic help for women in this extremely difficult time of their lives.

II. Gestagens

The Mechanism of Effect of Gestagens

H. P. Zahradnik

Abstract

The biological effects of progesterone depend on the presence of specific cytoplasmatic receptors. The synthesis of these receptors requires the presence of estrogens. In addition progesterone is involved in the metabolism of estrogens in target tissue. Progesterone stimulates the activity of the 17β-ol-dehydrogenase.

The present study follows the steroid and prostaglandin levels (PGF_{2a}, 13, 14-dihydro-15keto-PGF_{2a} and PGE_2) in the menstrual blood in normal women and in patients suffering from dysmenorrhea. Intrauterine application of progesterone results in a significant decrease of the PGF_{2a}-levels coinciding with improvement of the clinical symptoms. The results obtained present evidence that progesterone inhibits the estrogen dependent uterine prostaglandin synthesis.

When a method was found to prove that receptors were present in the uterus, a target organ for progesterone, the somewhat contradictory effects of this substance on the uterus became a bit more susceptible to explanation. For example, experiments in the guinea pig have shown that the number of progesterone receptors in the uterus varies in relation to the overall hormonal picture during the cycle. Progesterone plasma level and progesterone receptor concentration per uterine cell show cyclic changes, which, however, do not coincide. These observations seem to indicate that the amount of progesterone available around the time of ovulation is physiologically significant if the progesterone receptors are to play their obligatory role in implantation.

Under certain conditions progesterone does not exert biologic effects unless estrogens have been administered beforehand, while in other cases progesterone may even antagonize the estrogen effect. In the chicken oviduct, for instance, estrogens followed by progesterone will stimulate the production of a certain protein, avidin. In vivo, this avidin is detected about 10 h after progesterone administration, and in vitro it is already present after 6 h. During this time period one finds no alterations in or stimulation of the synthesis of other proteins.

O'Malley (1) was able to prove that the resynthesis of mRNA, the precondition for avidin formation, is dependent on progesterone. He isolated the RNA from the ribosome fraction of the oviduct of progesterone-stimulated chickens, and thanks to his highly specific method he was able to obtain pure mRNA. Using this highly purified mRNA O'Mally (1) and his group succeeded in synthesizing avidin in vitro. They were also able to show that an mRNA fraction obtained from oviducts that had been pretreated with estrogen only, brought about the synthesis of another specific protein, ovalbumin, in vitro.

What this means is that neither estrogens in the absence of progesterone can induce avidin synthesis, nor can progesterone induce avidin synthesis before the tubular glandular cells have been differentiated by means of estrogens. It seems that estrogens either aid the formation of the receptors for progesterone in these cells, or that they trigger the corresponding nuclear mechanisms that lead to avidin synthesis. Perhaps they do both.

That estrogens increase the number of estrogen receptors in endometrium and myometrium is not surprising. It is remarkable, though, that progesterone by itself will lead to a rise in the number of estrogen receptors in the endometrium but not in the myometrium. Now if progesterone is administered together with estradiol, the progesterone will suppress this estrogen-dependent increase in receptor number. The relative concentrations of receptors in endometrium and myometrium are, remarkably, very much the same in this progesterone-estrogen model as they are under physiologic conditions. The antagonistic effect that progesterone has on the induction of estradiol receptors in myometrium, may serve to aid blastocyst implantation by lowering the estrogen-dependent sensitivity of the myometrium to catecholamines and prostaglandin F_{2a} (PGF_{2a}). This mutual influence on the prepregnant uterus of progesterone and estrogens on the one hand, and steroids and prostaglandins on the other, was the subject of the in vivo investigation discussed below.

In eight women, five with dysmenorrhea and three with eumenorrhea, we determined by radioimmunologic assay the menstrual blood content of PGF_{2a}, 13,14-dihydro-15-keto-PGF_{2a} (DHK-PGF_{2a}), PGE_2, estradiol, estrone, estriol, and progesterone. After two untreated cycles we placed a progesterone-dispensing device in the uterus of the dysmenorrheic women. The above endocrine parameters were then determined after 1, 2, 6, and 12 cycles.

The columns in Figure 1, like those of the other figures, represent the percentage difference found for the above parameters with respect to the corresponding control

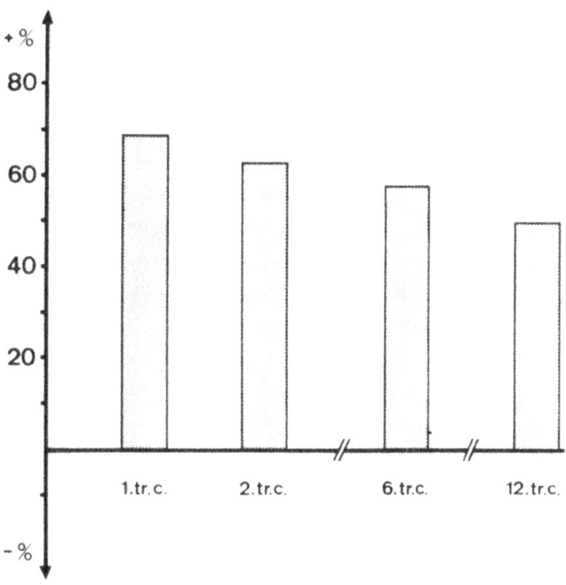

Fig. 1. Progesteronexcretion per menstruation (%-diff. to dysmenorrhoic cycles)

cycle. Reference point is the sum of the largest positive and negative difference between the levels measured before and after insertion of the intrauterine device. As you can see from this figure, *progesterone concentration* in the menstrual blood was considerably higher 1 month following the insertion of the device. Thereafter the percentage difference between treated (tr. c. in Figs. 1–6) and untreated cycles decreased steadily over the 12-month observation period.

In Figure 2 we see that immediately after insertion of the progesterone device, the *estradiol content* of menstrual blood increased by 34%. After 12 cycles the percentage difference to the control cycles was down to about 15%. After insertion of the intrauterine device the *estron level* increased by about 25%–30% over the dysmenorrheic control level, as you can see from Figure 3. These values remained about the same over the entire 1-year observation period. The *levels of estriol excretion* showed no differences between eumenorrheic and dysmenorrheic patients with or without progesterone treatment.

Figure 4 shows that PGF_{2a}, which we know correlates with pain in general and dysmenorrheic symptoms in particular, was significantly higher in the dysmenorrheic controls than in the patients with eumenorrhea. After the intrauterine device was inserted, which gave a daily dosage of 65 μg progesterone, the PGF_{2a} excretion fell sharply after two menstrual cycles. The same time period also brought a clear improvement in the clinical picture of dysmenorrhea.

Similarly, the excretion of the stable metabolite DHK-PGF_{2a} fell significantly after the intrauterine device (Progestasert) was inserted (Fig. 5). This indicates that the synthesis of PGF_{2a} is inhibited by progesterone.

Figure 6 shows that the menstrual excretion of PGE_2 increased under progesterone influence by about the same percentage as PGF_{2a} decreased. After the first cycle we found an increase of about 50% as against the control cycles in dysmenorrheic patients. This increase held over the entire 12-month period of observation.

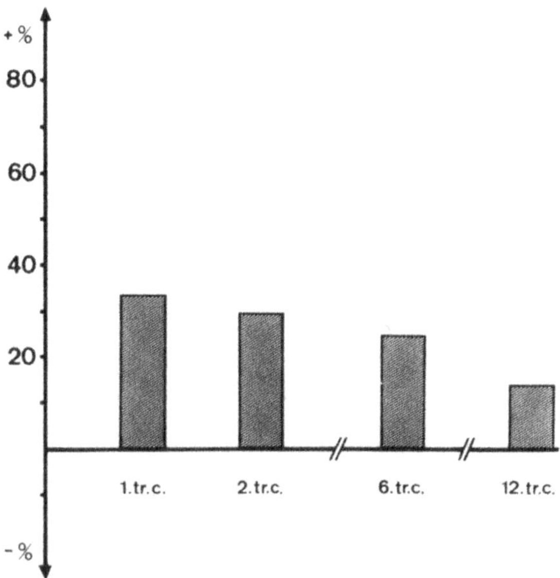

Fig. 2. Estradiolexcretion per menstruation (%-diff. to dysmenorrhoic cycles)

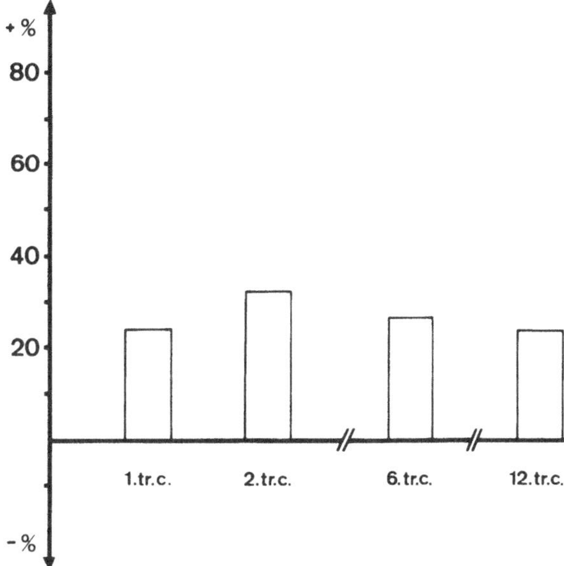

Fig. 3. Estronexcretion per menstruation (%-diff. to dysmenorrhoic cycles)

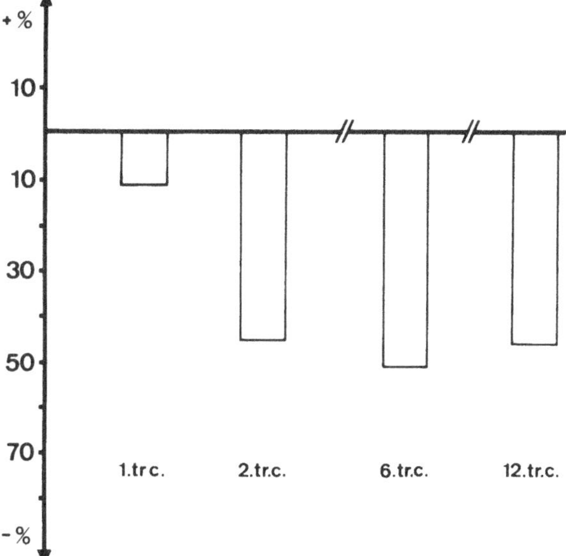

Fig. 4. Prostaglandin-F_{2a}-excretion per menstruation (%-diff. to dysmenorrhoic cycles)

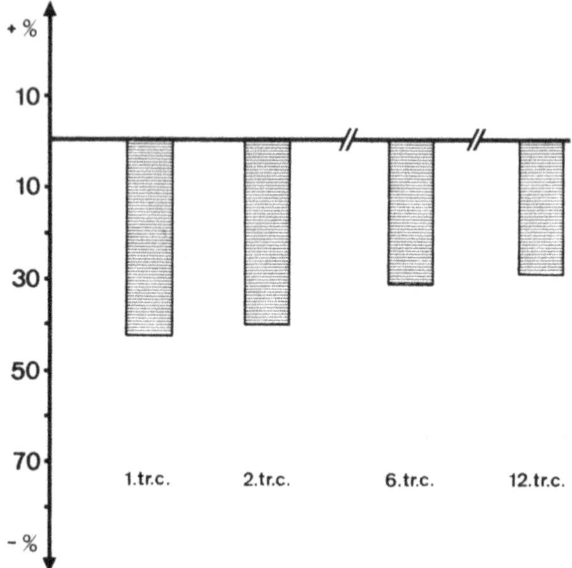

Fig. 5. 13,14-Dihydro-15-Keto-PGF_{2a}-excretion per menstruation (%-diff. to dysmenorrhoic cycles)

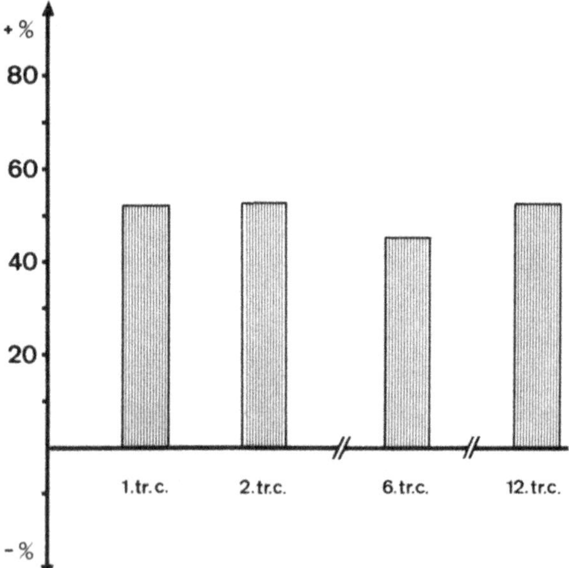

Fig. 6. Prostaglandin-E_2-excretion per menstruation (%-diff. to dysmenorrhoic cycles)

Thus it is quite clear that progesterone administered intrauterinely brings about a reduction in uterine PGF_{2a}. Also, progesterone obviously increases the rate of metabolism of estradiol to estron considerably, i.e., it stimulates 17β-ol-dehydrogenase activity. Our results also indicate that progesterone has some influence on the formation of estrogen receptors, which may be causally linked to the higher menstrual blood levels of estradiol we observed. It would be presumptuous on the basis of our analysis to speculate on the subcellular mechanisms involved; yet the reduction of cytoplasmic receptor formation that has often been seen with progesterone would go far towards explaining our results, as would changes in the receptor, area such as those described by Hsueh in Nature in 1975 (2). This alteration in the area of the estrogen receptors is surely also responsible for the considerable reduction in PGF_{2a} formation and for the fact that PGE_2 is found in greater amounts following intrauterine application of progesterone.

What we are seeing, in other words, are two biologic counterregulation mechanisms on the uterine level that have their counterparts on the receptor level. PGF_{2a} is antagonized in its biologic activity by PGE_2, while progesterone inhibits the effect of estradiol. That this hypothesis is also valid for the pregnant uterus has been shown by other investigations, which, however, we will not go into at this point.

References

O'Malley BW, McGuire WL, Kohler PO, Korenman SG (1969), Rec Prog Horm Res 25:105

Hsueh M (1975) Nature 254:339

Pharmacokinetic and Pharmacodynamic Effects of Small Doses of Norethisterone Released from Vaginal Rings Continuously During 90 Days

Britt-Marie Landgren, Elisabeth Johannisson, Britt Masironi and E. Diczfalusy

Abstract

The pharmacokinetic and pharmacodynamic effects of norethisterone (17 α-ethinyl-17ß-hydroxy-4-estren-3-one; NET) released continuously from vaginal devices at a rate of 50 μg/24 h and 200 μg/24 h, respectively, were investigated during a 90-day period in two groups of seven women each with regular menstrual periods. Blood samples were taken three times weekly (mondays, wednesdays, and fridays) during a control cycle and during the entire study period for the estimation of estradiol, progesterone, and NET levels, and hourly blood samples were collected throughout a 24-hour period after 6 weeks of exposure to assess the short term variation in NET levels. In addition, an endometrial biopsy was taken on days 21-23 of the control cycle and after 6 and 10 weeks with the device in situ.

There was little, if any, initial "burst" effect on the plasma levels of NET following the insertion of the devices, and a statistically significant linear relationship was found in each case when the NET levels were plotted against the 90 days of exposure. However, the average slope of the regression lines with the 50 μg releasing devices (-0.44) was significantly flatter than that with the 200 μg releasing devices (-1.63). Also the mean plasma level in the former group 283 pg/ml) was significantly lower than that (666 pg/ml) observed in the latter group. The average daily decline in plasma NET levels was 0.16% and 0.24%, respectively, and the 95% confidence limits of the hour-to-hour variation in NET levels were at 92% and 106%.

All control cycles were of normal length (26-35 days) and exhibited a normal luteal activity. There were a total of 16 cycles with normal luteal function among the seven subjects with 50 μg releasing devices; each of them had a minimum of two such cycles. However, 11 of these 16 cycles were of abnormal length, indicating a significant effect of NET on the follicular phase. Only nine cycles were found with a normal luteal activity in the group with the devices releasing 200 μg/24 h; three of the seven subjects in this group had anovulation throughout the entire study and two of them had prolonged anovulatory periods alternating with normal cycles.

Thirteen of the 14 biopsies taken with the 50 μg releasing devices in situ after 6 and 10 weeks of exposure showed normal cyclic changes, whereas 11 of the 14 biopsies taken in the group exposed to a release rate of 200 μg/24 h indicated predecidual and atrophic changes. These were associated with a significantly diminished number of endometrial glands per microscopic field and with a significant increase in the number of days will bleeding and spotting. Hence, the increased frequency of intermenstrual bleeding may be associated with the predecidual and atrophic changes induced by the administration of NET.

No correlation was found between the individual plasma levels of NET and the following parameters: weight, height and ponderal index of the subjects, individual release rate of NET from the devices, effects on ovarian function, the appearance of the endometrium, and the number of days with bleeding and spotting.

It is concluded that the vaginal rings studied provide a near zero order release of NET and represent an ideal experimental model for the study of the effects of constant blood levels of contraceptive steroids.

Since the pioneering studies of Mishell and co-workers (Mishell et al. 1970; Mishell and Lumkin 1970), a considerable number of investigations have been reported on the vaginal administration of contraceptive steroids in relatively high, ovulation-inhibiting doses (Mishell et al. 1972; Hiroi et al. 1975; Thiery et al. 1976; Victor and Johansson 1976a, b; Mishell and Lumkin 1975; Stanczyk et al. 1975; Mishell et al. 1977; Victor et al. 1975; Viinikka et al. 1975; Johansson et al. 1975; Akinla et al. 1976; Henzl et al. 1973; Mishell et al. 1978; Victor et al. 1977). Vaginal devices releasing constant amounts of steroids appear to represent an attractive approach to contraception, since they are self-administered, long-acting, their presence in the vagina seems to cause little, if any, inconvenience, and the exposure to the contraceptive steroid can be rapidly discontinued at any time by easy self-removal. In addition, vaginal administration of steroids results in an excellent absorption and avoids the "first passage effect" through the liver.

There is also another useful application for vaginal devices, namely to provide suitable pharmacodynamic models for the assessment of the likely effects of contraceptive steroids released at near zero order, e.g., from biodegradable implants. Such implants aim at eliminating the major fluctuation in the plasma levels of orally administered steroids, hoping that in this way some of the side effects can be reduced. It is also conceivable that a zero order release may considerably reduce the amount of steroid needed for contraception and thus the total steroid load to be metabolized. Were it possible to achieve constant plasma levels of contraceptive steroids following their release from vaginal devices, the above propositions could be tested experimentally.

Hence, the aim of the present study was to provide an in-depth assessment of the pharmacokinetic and pharmacodynamic effects of NET released continuously over a period of 90 days from vaginal rings in constant small doses, which are not expected to inhibit ovulation in the majority of subjects studied.

Material and Methods

Vaginal devices

These were fabricated by the Battelle Memorial Institute, Pacific Northwest Division (Richland, Wa., USA) under a contract from the WHO. The devices were toroidal-shaped, with an outside diameter of 55.6 mm and a cross-section of 9.5 mm. They were fabricated entirely of Dow-Corning 382 Silastic Medical Grade Elastomer, a polydimethylsiloxane silicone rubber. The details of fabrication and in vitro testing were reported elsewhere (Burton et al. 1978). The expected release rates were 50 μg/24 h and 200 μg/24 h.

Clinical Material

Fourteen apparently healthy women (aged 21 to 36 years) with a minimum of 3-months' history of regular menstrual periods (26 to 35 days) volunteered for this investigation. For a minimum of 3 months prior to the start of the study they did not use any steroidal contraceptives, or intrauterine devices. Their hematologic status was carefully checked before, during, and immediately after the completion of the study. The volunteers were allocated randomly to the two types of devices.

Plan of Study

Following an untreated (control) cycle, in which blood samples of 10 ml were withdrawn three times weekly (mondays, wednesdays and fridays between 08.00 and 10.00 h) and an endometrial biopsy specimen was taken on day 21-23 of the cycle (using a Randall curette without dilatation of the cervix and without anesthesia), the vaginal devices were inserted on the 5th day of the next cycle and were left in situ for an average of 90 (range: 86-94) days. During the entire period of exposure to the devices, blood samples were withdrawn at the same frequency and in the same way as during the control cycle, and two endometrial biopsy specimens were taken during the 6th and 10th week of exposure. In connection with the endometrial biopsy, additional blood samples of 5.0 ml each were taken at hourly intervals during a period of 24 h (25 samples) in order to assess the hour-to-hour variation in NET levels. Hence, the total volume of blood withdrawn from each volunteer during a period of 120 days amounted to 545 ml.

Because of the uncertainly as to the contraceptive efficacy of the small doses of NET released, the couples were advised to use a specified mechanical method of contraception (condoms). No pregnancies occurred.

The total exposure of the seven women with the 50 μg releasing devices was 634 days and of the seven women with the 200 μg releasing devices 631 days. Upon termination of the study, the vaginal devices were removed and sent to the Battelle Memorial Institute for the estimation of the residual content (i.e., release rate) of NET.

Steroid Assays

The blood samples obtained by vacutainers were centrifuged within 20 min, frozen and stored at -20°C until processed. In each plasma sample, progesterone and estradiol were estimated by the radioimmunoassay method described by Aso et al. (1975) and NET by the method of Bedolla-Tovar et al. (1978).

Histologic Examination

The endometrial specimens were immediately fixed in Bouin's solution, embedded in paraffin, sectioned, and stained with hematoxylin-eosin. The biopsies were evaluated with regard to the presence of irregular secretory changes, predecidual reaction and atrophic changes in the glands. In addition, the average number of endometrial glands per microscopic field (x400) was quantitated. Only glands having a perpendicular section were included in this study. For dating of the biopsies taken in the control cycle the criteria described by Noyes et al. (1950) were employed.

Classification of Treatment Cycles

This was based on the outcome of a series of recent studies conducted in Stockholm, in which daily hormone assays were carried out in the blood of 72 normally menstruating women throughout a complete cycle. The (geometric) mean length of these cycles was 29.7 days, with 90% tolerance limits (covering 90% of the population with 90% probability) at 26 and 35 days (Landgren, in manuscript). Ninetyeight percent of these women had a preovulatory estradiol surge between 100 and 700 pg/ml and a luteal estradiol maximum between 100 and 500 pg/ml. Furthermore, a study of these 72 cycles together with the data of 34 cycles published previously (Guerrero et al. 1976; Aedo et al. 1976) revealed that 94% of these 104 cycles exhibited a plasma progesterone level of 5.0 ng/ml or more for a minimum of 5 days. Hence, in the present study a cycle is classified as normal only if it fulfills the above criteria with regard to cycle length and peripheral estradiol and progesterone levels. Furthermore, anovulation with no luteal function is characterized by constant progesterone levels (as a rule below 1.5 ng/ml) found in periods in which the mean value of the second part of the "cycle" (i.e., observation period) does not differ significantly from that found in the first part. Finally, "depressed" luteal function is characterized by significantly, but insufficiently elevated plasma progesterone levels (less than 5.0 ng/ml for 5 days) during the second half of the observation period.

Results

Pharmacokinetic Observations

Figures 1 and 2 present some representative examples of the study and indicate at the same time some of the problems posed by near zero order delivery systems.
The data of Fig. 1 indicate rather high plasma levels of NET (around 1000 pg/ml) associated with normal ovulatory cycles as evidenced by the estradiol and progesterone levels. In the other subject (Fig. 2), the NET levels were invariably below 400 pg/ml; nevertheless, ovulation was inhibited throughout the entire duration of the study. The data of Figs. 1 and 2 also indicate the virtual lack of any initial "burst" effect in NET levels following the insertion of the devices.

A statistically significant linear relationship was found in each of the 14 cases when the plasma levels of NET were plotted against the 90 days of exposure. The slope values and Y-intercepts of the individual regression lines, together with the estimated release rates of the individual devices, are indicated in Table 1.
Although there is some overlap between the two groups concerning both the slope- and Y-intercept values, both differences are statistically significant ($p < 0.05$ and p

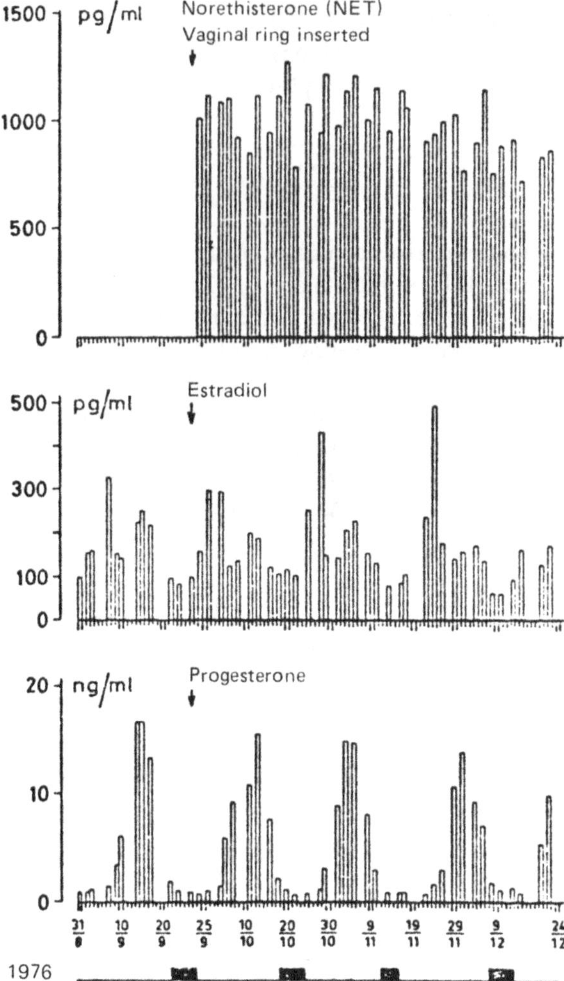

Fig. 1. Systemic plasma levels of NET, estradiol, and progesterone in a subject before and during continuous exposure to norethisterone released from a vaginal device at a rate of 360 μg/24 h. Periods of bleeding are indicated by filled bars

< 0.01, respectively). Hence devices releasing 200 μg/24 h give rise to higher plasma levels of NET than those releasing 50 μg/24 h. Furthermore, the plasma levels of NET decline significantly faster in the former group. It follows from the regression coefficients that during the 90 days of study, the average plasma level of NET in the group with the 50 μg releasing devices declined by 14%, and in the group with 200 μg releasing devices by 22%, corresponding to a daily decrease of 0.16% in the former and 0.24% in the latter group.

Moreover, the data of Table 1 also indicate a mean release rate of 46 μg/24 h (with 95% confidence limits at 35 and 59 μg/24 h) for the devices expected to release 50 μg/24 h, and a release rate of 163 (86–310) for the 200 μg releasing rings, with the proviso that if the outlier value of 40 μg/24 h in the high release group is disregarded, a more likely mean release rate of 206 (144–295) μg/24 h may be obtained.

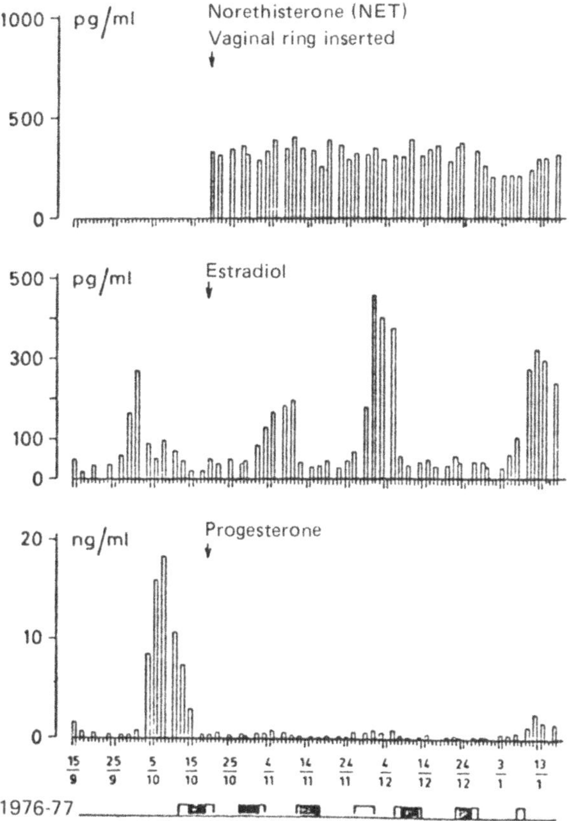

Fig. 2. Systemic plasma levels of NET, estradiol, and progesterone in a subject before and during continuous exposure to NET released from a vaginal device at a rate of 200 μg/24 h. Periods of spotting are indicated by open bars, those of bleeding by filled bars

There was no correlation between the release rates and plasma levels, neither in the 50 μg release group (r=0.183), nor in the 200 μg-release group (r=0.569; limit of significance at r=0.811). The exclusion of the outlier single value (40 μg/24 h) in the high release group did not change the significance of the correlation. Furthermore, the poor correlation was not improved after the values were adjusted for differences in weight, height and ponderal index.

Does the vaginal device provide constant plasma levels of NET around the clock, or are there major differences between periods of activity and rest? In order to answer this question, the hour-to-hour changes in plasma levels were assessed during a 24-h period, as shown in Table 2.

The data of Table 2 indicate a great uniformity of plasma NET levels; in none of the 14 subjects did the slope of the regression line – calculated from the 25 estimates around the clock (not shown in Table 2) – differ significantly from zero, indicating that during this period a true zero order release was achieved.

Table 1. Individual regression coefficients (slope and intercept values on the Y-axis)[a]

Expected Release Rate	Subject	Regression coefficient (slope)	Intercept (pg/ml)	Estimated release rate[b] (μg/24 h)
	M.K.	-0.11	280	58
	L.L.	-0.22	199	46
	M.W.	-0.28	265	40
50 μg/24 h	W.M.	-0.42	149	53
	M.G.	-0.74	430	50
	B.S.	-0.76	327	57
	A.W.	-2.12	466	26
Mean (geom.)		-0.44[c]	283	46
95% confidence limits		-0.18-1.08	194 - 411	35 - 59
	A.K.	-0.71	525	130
	M.H.	-0.98	699	240
	C.L.	-1.01	361	200
200 μg/24 h	B.S.	-1.32	535	40[d]
	K.N.	-2.86	1115	170
	B.A.	-2.99	727	200
	B.W.	-3.84	1013	360
Mean (geom.)		-1.63[e]	666	163
95% confidence limits		-0.88-3.00	463 - 959	86 - 310

[a] Calculated from the regression lines of the plasma levels of NET (pg/ml) measured during 90 days in 14 subjects with vaginal devices releasing NET at a rate of 50 μg/24 h and 200 μg/24 h, respectively. The individual release rates estimated from the amount of NET recovered from the devices at the end of the study are also indicated

[b] These estimations were kindly carried out by the Battelle Memorial Institute, Pacific Northwest Division, Richland, Wa. USA

[c] Arithmetic mean value : -0.66 ± 0.25

[d] If this outlier is disregarded, a mean release rate of 206 (144-295) μg/24 h is obtained

[e] Arithmetic mean value : -1.96 ± 0.47

Table 2. Variation in plasma levels of NET in 25 peripheral blood specimens obtained at hourly intervals from 14 women with vaginal devices releasing NET at different rates

Release Rate	Mean (geom.) (pg/ml)	95% Confidence limits	Confidence limits (%)
	134	125 – 144	87 – 107
	182	174 – 189	92 – 104
50 μg	222	198 – 250	89 – 113
(7 subjects)	234	224 – 245	91 – 105
	234	217 – 252	93 – 108
	262	242 – 283	92 – 108
	363	347 – 379	92 – 104
	290	272 – 309	94 – 107
	431	411 – 452	95 – 105
200 μg	483	461 – 507	95 – 105
(7 subjects)	561	538 – 584	96 – 104
	578	553 – 605	91 – 105
	778	717 – 844	92 – 108
	831	796 – 868	92 – 104

Pharmacodynamic Events

The events observed with the two types of release are summarized in Table 3.

Table 3. Incidence of pharmacodynamic events during 3 subsequent cycles in 14 subjects with vaginal devices releasing NET at different rates

Events	Release Rate/24 h	
	50 μg (7 subjects)	200 μg (7 subjects)
Ovulatory cycle	16	9
Cyclic follicular Activity with depressed Luteal function	3	9
Ovarian supression	1	4
Total	20	22

A normal cycle length and a normal luteal phase were found in all 14 control cycles (not shown in Table 3). Furthermore, there were a total of 16 cycles with normal luteal function among the seven subjects during exposure to the 50 μg releasing devices; each of them exhibited a minimum of two such cycles. However, 6 of these 16 cycles were shorter than 26 days, and 5 of them were longer than 35 days. Moreover, the luteal function was significantly depressed in three cycles and absent in one.

Only nine cycles were found with a normal luteal function in the group exposed to devices releasing 200 μg/24 h; three of the seven subjects in this group experienced anovulation throughout the entire study, two had regular ovulatory cycles, and in two of them prolonged anovulatory periods alternated with normal cycles. Moreover, a total of nine instances with cyclic follicular activity without any luteal activity were observed in this group.

Another effect of the small doses of NET during exposure to a device releasing 50 μg/24 h of NET is indicated by the data of Fig. 3;

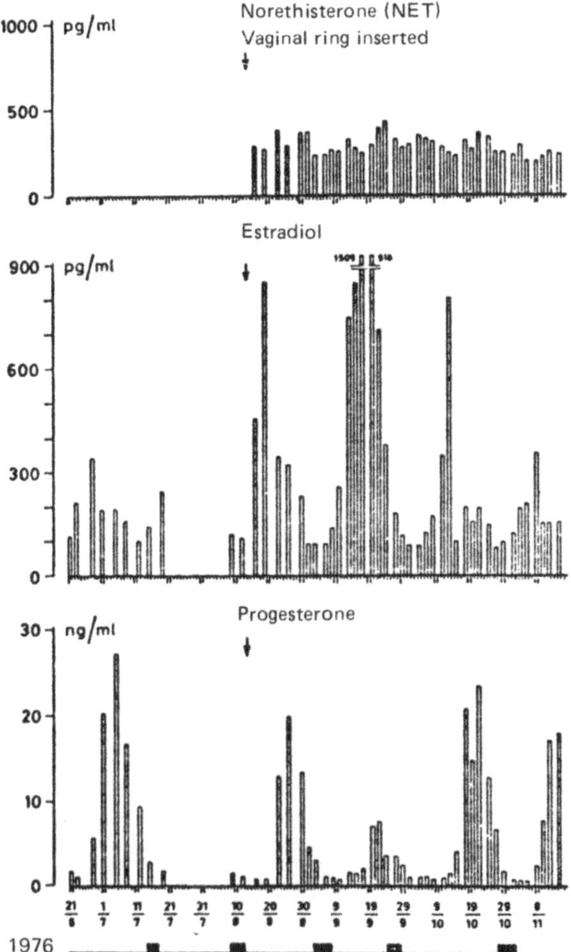

Fig. 3. Systemic levels of NET, estradiol, and progesterone in a subject before and during continuous exposure to NET released from a vaginal device at a rate of 57 μg/24 h. Periods ob bleeding are indicated by filled bars

in this case, normal luteal phases were associated with abnormally elevated (> 800 pg/ml) preovulatory estradiol surges in each of three subsequent treatment cycles. A similarly elevated estradiol surge was also seen in another subjects with a device releasing 200 μg/24 h. As indicated before, in our normal material only one of 72 normally menstruating women exhibited a plasma estradiol level exceeding 700 pg/ml.

Endometrial Morphology

The results of the histologic examination are shown in Table 4.

All 14 pretreatment (control) biopsies and 13 out of 14 biopsies obtained in the seven subjects during exposure to the devices releasing NET at a rate of 50 μg/24 h indicated normal cyclic changes. On the other hand, only 3 of the 14 biopsies obtained in the subjects with the 200 μg releasing devices in situ showed a normal cyclic change; six exhibited a predecidual reaction and five showed atrophic changes. Furthermore, there was a close agreement between the results of the histologic dating of the endometrium and the classification of the day of the cycle on the basis of the plasma steroid levels in 14 of the double-blind comparisons made during the pretreatment period, in 13 out of 14 comparisons made during exposure to the devices releasing 50 μg/24 h, but only in seven of the comparisons made during exposure to 200 μg releasing devices. In the latter group, predecidual, or atrophic changes of the endometrium were found sometimes in the presence of hormonal changes indicating a normal ovulatory cycle, and sometimes in anovulatory cycles with no signs of cyclic estradiol or progesterone secretion.

There was no correlation between the plasma levels of NET and the histologic appearance of the endometrium; relatively high (> 1000 pg/ml), or relatively low (< 400 pg/ml) plasma levels of NET were associated sometimes with a normal and sometimes with an atrophic endometrium.

In an attempt to quantitate the predecidual and atrophic changes of the endometrium, the number of glands per microscopic field was estimated. The results are presented in Table 5.

Table 4. Histologic assessment of the endometrial biopsy specimens obtained from 14 women during a pretreatment (control) cycle and during the 6th and 10th week, respectively, with the NET releasing device in situ

Endometrial appearance	Release rate: 50 μg/24 h			Release rate: 200 μg/24 h		
	Pretreatment	6 weeks	10 weeks	Pretreatment	6 weeks	10 weeks
Normal cyclic	7	6	7	7	1	2
Predecidual	0	1	0	0	4	2
Atrophic	0	0	0	0	2	3

Table 5. Number (± SD) of endometrial glands per microscopic field in 14 women before and 6 and 10 weeks after the insertion of vaginal rings releasing NET at different rates

Number of subjects	Release rate (μg/24 h)	Control	6 weeks	10 weeks
7	50	3.1 ± 0.8	3.1 ± 1.0	3.9 ± 2.0
7	200	3.2 ± 0.8	1.6 ± 0.8	1.0 ± 0.7

Table 6. Bleeding and spotting in 14 women with vaginal devices releasing NET at different rates

Release rate/24 h	No. of subjects	Days with bleeding and spotting	Days of exposure	Average per subject (± SD)	Significance
50 μg	7	103	631	14.7 ± 3.0	$p < 0.05$
200 μg	7	179	634	25.6 ± 9.6	

The data of Table 5 indicate that there was a significant decrease in the number of glands during exposure to the devices releasing 200 μg of NET, whereas devices releasing 50 μg/24 h had no effect.

Bleeding and Spotting

As indicated by the data of Table 6, the number of days with bleeding and spotting was significantly higher in the subjects with devices releasing 200 μg/24 h than in those with the devices releasing 50 μg/24 h.

Discussion

Perusal of the literature on vaginal rings releasing contraceptive steroids reveals that most, if not all studies published address themselves to devices which release ovulation inhibiting, high doses of medroxyprogesterone acetate (6α-methyl-17α-acetoxy-4-pregnene-3,20-dione) (Mishell et al. 1970; Mishell and Lumkin 1970; Mishell et al. 1972; Hiroi et al. 1975; Thiery et al. 1976; Victor and Johansson 1976a) (Viinikka et al. 1975; Johansson et al. 1975; Akinla et al. 1976), norgestrel [(±) 13β-ethyl-17α-ethinyl-17β-hydroxy-4-gonen-3-one] (Mishell and Lumkin 1975; Stanczyk et al. 1975; Mishell et al. 1977; Victor et al. 1975; Victor and Johansson 1976b), or ethylnorgestrienone (R-2323) [(±) 13β-ethyl-17α-ethinyl-17β-hydroxy-4,9 (10)-gonatrien-3-one]. In addition, a few papers deal with systems releasing chlormadinone acetate [6-chloro-17α-acetoxypregna-4,6-diene-3,20-dione] (Henzl et al. 1973), or a combination of norgestrel with estradiol (Mishell et al. 1978), or estradiol benzoate (Victor et al. 1977). It would appear that only one previous paper (Mishell and Lumkin 1975) reports on devices releasing NET. In this study, vaginal rings containing 100 and 200 mg of NET were used; a high incidence of bleeding was observed and ovulation occurred in about one-fourth of the treatment cycles. The estimated release rate was higher than 1500 μg/ 24 h.

The present study was aimed at exploring the pharmacokinetic and pharmacodynamic properties of NET when released at a constant rate from vaginal devices in such small amounts which are unlikely to invariably inhibit ovulation. The data herein reported indicate that the devices do release almost constant amounts of NET for 90 days, or longer, as evidenced by the linear relationship between plasma levels of NET and the days of exposure during the entire span of the study. Furthermore, the data obtained on the hour-to-hour variation of plasma levels indicate that by the use of the devices studied it is possible to obtain a true zero order release of NET. Hence, these devices represent an ideal experimental tool for studies on the effect of constant plasma levels of contraceptive steroids.

The data reported here also indicate that with devices of this type there is little, if any, initial "burst" effect. Such an effect usually manifests itself in considerably higher plasma levels during the 1st few days of the study (e.g. Hiroi et al. 1975), or longer (e.g. Benagiano 1977), and represents a considerable drawback, since it greatly complicates the evaluation of the observed pharmacodynamic effects.

Statistically valid linear regression lines were obtained for the plasma levels of NET throughout the entire experiment in each subject. The mean regression coefficients obtained in the two goups (-0.44 and -1.63, respectively) indicate that the average daily decline in plasma NET levels was only 0.16% and 0.24%, respectively, and that thus a near zero order release of NET was achieved during the 90 days of the study. Although there was a considerable overlap and great individual variation in the plasma levels of NET, the (geometric) mean plasma level in subjects with the 200 μg releasing devices was significantly higher than that in the group with 50 μg releasing rings. On the other hand, and in agreement with a previous report on norgestrel-releasing vaginal rings (Victor and Johansson 1976b) no correlation was found between the individual release rates from the devices and the plasma levels of NET in the same subjects. Furthermore, the plasma levels of NET were not correlated to height, weight and ponderal index of the subjects, or to their effect on ovarian function, endometrial morphology, or the frequency of bleeding and spotting. It would seem, therefore, that other factors, such as specific binding proteins in the circulation and in the endometrium may require a careful analysis in this respect.

In a recent study on the pharmacodynamic effects of the NET minipill (300 μg daily), we have assessed the day-to-day changes in the plasma levels of progesterone and estradiol in 42 women during a control cycle and during the second treatment cycle. Using the same criteria for assessment as in the present study, a normal luteal function was found in 40% of the subjects, 20% had a significantly depressed luteal activity, and 40% exhibited no luteal function whatsoever (Landgren et al., in manuscript). In the present study, three of the seven subjects with the devices releasing NET at a rate of 200 μg/24 h had anovulation throughout the entire study period, and two had anovulatory periods alternating with cycles with normal luteal function. Hence, it would appear – as a first approximation – that the ovarian effect of a 300 μg NET minipill is roughly equivalent to that of 200 μg of NET released continuously from a vaginal device during 24 h.

The pharmacodynamic effect of the devices releasing 50 μg/24 h was different. The luteal function was significantly depressed or absent in 4 cycles, but there were 16 cycles with normal luteal function and a minimum of 2 such cycles in each subject. However, 11 of the 16 cycles with normal luteal function had an abnormal length. Hence, the first effect of NET when released in doses insufficient to inhibit ovulation is directed towards the (incompletely comprehended) factors regulating cycle length in general and the length of the follicular phase, in particular. This suggests that it might be rewarding to conduct an in-depth study on the plasma levels of a variety of pituitary hormones during the follicular phase of cycles with a normal luteal phase in women exposed to similarly small doses of progestogens which are insufficient to inhibit ovulation.

Whereas in 13 of the 14 endometrial biopsies taken in the group with the 50 μg-releasing devices normal cyclic changes were observed, 11 of the 14 biopsies taken in the group with the 200 μg releasing devices showed predecidual or atrophic changes. These changes were associated with a significant decrease in the number of endometrial glands and with a significant increase in the frequency of bleeding and spotting. On the other hand, no correlation was found between the endometrial changes observed and the circulating levels of NET, estradiol, or progesterone. This seems to speak against the existence of a simple relationship between peripheral hormone levels (exogenous or endogenous) and intermenstrual bleeding in women using progestogen minipills for contraception. Thus, the present study points to the necessity of conducting a series of studies at the endometrial level in order to identify the major factors responsible for intermenstrual bleeding in minipill users. It is felt that there is little prospect for progress in this field, unless the mechanism of intermenstrual bleeding is better understood.

Finally, it should be emphasized that the vaginal devices were excellently tolerated by all subjects. This fact, the safety data obtained in animal toxicologic studies and the bleeding patterns observed seem to justify an exploratory study of the contraceptive efficacy of vaginal rings releasing NET at a rate of approximately 50 μg/24 h. The frequency of bleeding and spotting seen with the rings releasing NET at a rate of 200 μg/24 h seems to militate against the use of devices with this and higher release rates, unless a possible combination with an estrogen is considered (e.g. Mishell et al. 1978; Victor et al. 1977). However, such an approach will necessitate more information on the safety of estrogens released continuously into the vagina.

Acknowledgement

This investigation received financial support from the World Health Organization; Geneva, and from the Ford Foundation, New York.

References

Aedo AR, Landgren BM, Cekan Z, Diczfalusy E: (1976) Studies on the pattern of circulating steroids in the normal menstrual cycle. 2. Levels of 20α-dihydroprogesterone, 17-hydroxyprogesterone and 17-hydroxypregnenolone and the assessment of their value for ovulation prediction. Acta Endocrinol (Kbh) 82:600

Akinla O, Lähteenmäki P, Jackanicz TM (1976) Intravaginal contraception with the synthetic progestin, R2323, Contraception 14:671

Aso T, Guerrero R, Cekan Z, Diczfalusy E (1975) A rapid 5-hour radioimmunoassay of progesterone and oestradiol in human plasma. Clin Endrocrinol (Oxf) 4:173

Bedolla-Tovar N, Rahman SA, Cekan S, Diczfalusy E (1978) Assessment of the specificity of norethisterone radioimmunoassays. J Steriod Biochem 9:561

Benagiano G (1977) Long-acting systemic contraceptives. In: Diczfalusy E (ed) Regulation of human fertility. WHO Symposium, Moscow 1976. Scriptor, Copenhagen, pp 323–360

Burton FG, Skiens WE, Gordon NR, Veal JT, Kalkwarf DR, Duncan GW (1978) Fabrication and testing of vaginal contraceptive devices designed for release of prespecified dose levels of steroids, Contraception 17:221

Guerrero R, Aso T, Brenner PF, Cekan Z, Landgren BM, Hagenfeldt K, Diczfalusy E (1976) Studies on the pattern of circulating steroids in the normal menstrual cycle. 1. Simultaneous assays of progesterone, pregnenolone, dehydroepiandrosterone, testosterone, dihydrotestosterone, androstenedione, oestradiol and oestrone. Acta Endocrinol (Kbh) 81:133

Henzl MR, Mishell DrJr, Giner Velazquez J, Leitch WE (1973) Am J Obstet Gynecol 115:101

Hiroi M, Stanczyk FZ, Goebelsmann U, Brenner PF, Lumkin ME, Mishell DRJr (1975) Radioimmunoassay of serum medroxyprogesterone acetate (Provera) in women following oral and intravaginal administration. Steroids 26:373

Johansson EDB, Luukkainen T, Vartiainen E, Victor A (1975) The effect of progestin R2323 released from vaginal rings on ovarian function. Contraception 12:299

Mishell DRJr, Lumkin ME (1970) Contraceptive effect of varying dosages of progestogen in silastic vaginal rings. Fertil Steril 21:99

Mishell DRJr, Lumkin M (1975) Initial clinical studies of intravaginal rings containing norethindrone and norgestrel. Contraception 12:253

Mishell DRJr, Talas M, Parlow AF, Moyer DL (1970) Contraception by means of a silastic vaginal ring impregnated with medroxyprogesterone acetate. Am J Obstet Gynecol 107:100

Mishell DRJr, Lumkin M, Stone S (1972) Inhibition of ovulation with cyclic use of progestogen-impregnated intravaginal devices. Am J Obstet Gynecol 114:927

Mishell DRJr, Roy S, Moore DE, Brenner PF, Page MA, Gentzschein E, Fisk PD (1977) Clinical performances and endocrine profiles with contraceptive vaginal rings containing d-norgestrel. Contraception 16:625

Mishell DRJr, Moore DE, Roy S, Brenner PF, Page MA (1978) Clinical performance and endocrine profiles with contraceptive vaginal rings containing a combination of estradiol and d-norgestrel. Am J Obstet Gynecol 130:55

Noyes RW, Hertig AT, Rock I (1950) Dating the endometrial biopsy. Fertil Steril 1:3

Stanczyk FZ, Hiroi M, Goebelsmann U, Brenner PF, Lumkin ME, Mishell DRJr (1975) Radioimmunoassay of serum d-norgestrel in women following oral and intravaginal administration. Contraception 12:279

Thiery M, Vandekerckhove D, Dhont M, Vermeulen A, Decoster JM (1976) The medroxyprogesterone acetate intravaginal silastic ring as a contraceptive device, Contraception 13:605

Victor A, Johansson EDB (1976a) Pharmacokinetic observations on medroxyprogesterone acetate administered orally and intravaginally. Contraception 14:319

Victor A, Johansson EDB (1976b) Plasma levels of d-norgestrel and ovarian function in women using intravaginal rings impregnated with dl-norgestrel for several cycles. Contraception 14:215

Victor A, Edqvist LE, Lindberg P, Elamsson K, Johansson EDB (1975) Peripheral plasma levels of d-norgestrel in women after oral administration of d-norgestrel and when using intravaginal rings impregnated with d-norgestrel. Contraception 12:261

Victor A, Nash HA, Jackanicz TM, Johansson EDB (1977) Collagen bands: A new vaginal delivery system for contraceptive steroids. Contraception 16:125

Viinikka L, Victor A, Jänne O, Raynaud J-P (1975) The plasma concentration of a synthetic progestin, R 2323, released from polysilastic vaginal rings. Contraception 12:309

Morphological Changes Induced by Exogenous Gestagens in Normal Human Endometrium

Gisela Dallenbach-Hellweg

Abstract

The synthetic gestagens used clinically differ both chemically and metabolically from natural progesterone. Most are derivatives of 17α-hydroxyprogesterone or 19-nor-testosterone. Although their gestogenic potencies vary considerably, most are more potent than natural progesterone. Norgestrel, which in its d-form is the most active gestogen known at present, is fifty times more active than progesterone.

The administration of progesterone alone during the proliferative phase depresses the maturation of Graafian follicles, arrests endometrial proliferation, and prevents ovulation. Even if given only for short periods, progesterone prolongs the menstrual cycle. When progesterone is discontinued a withdrawal bleeding occurs within a few days; when continued at low doses a breakthrough bleeding develops during therapy. If doses of 5–6mg are given daily for 4 weeks or longer, then the secretory change is abolished and the endometrium remains in a state of "arrested proliferation." With treatment beyond 6 weeks the arrested proliferation gives way to progressive atrophy of the glands and decidualization of the stroma. After three months of therapy, amounting to a total dose of about 500 mg progesterone, a typical appearing decidualized stroma develops with extreme or complete atrophy of the glands: "arrested secretion." If treatment is continued it may eventually lead to an irreversible atrophy with hyalinization of the stroma. The various synthetic gestagens differ both quantitatively and qualitatively in their action. The dosage required to produce a transformation of the endometrium varies from preparation to preparation. Moreover, some gestagens may affect mostly the stroma, others primarily the glands. Following therapy with derivatives of 19-nor-testosterone decidualization is more pronounced and the subsequent atrophy more extreme than with derivatives of progesterone. Since the glandular epithelial cells are more sensitive to progesterone and react to it earlier than the stromal cells do, the epithelium usually becomes refractory sooner to abnormal stimulation by gestogens, whereas the stroma begins to atrophy only after a prolonged decidualization.

Intrauterine contraceptive devices impregnated with progesterone promote focal decidualization sharply limited to the upper functionalis adjacent to the device. The striking histologic picture of a localized arrested secretion overlying a normally proliferating or secretory endometrium is characteristic.

In the endocervix gestagens induce adenomatous hyperplasia and reserve cell hyperplasia.

The synthetic gestagens used in contraceptives differ from natural progesterone both chemically and metabolically. Most are derivatives of 17α-hydroxyprogesterone or 19-nor-testosterone. Though the potencies of these substances vary, they are usually much more potent than their naturally occurring relatives. Norgestrel, the d-form of which is the strongest known gestagen, is fully 50 times more active than progesterone.

As might be expected, the effects of these synthetic gestagens are much more dose-dependent than those of natural progesterone, whereby the amount of substance administered is not as crucial as its potency. These effects will also vary with respect to the chemical structure of the gestagen used.

Progesterone given alone during the proliferation phase, leads in the *endometrium* to a delay in follicle maturation, proliferation and ovulation, and thus delays the entire cycle. When it is discontinued, withdrawal bleeding will occur within a few days; when continued at low doses, breakthrough bleeding develops. If doses of 5 to 6 mg are given daily for a period of 4 weeks or longer, secretory transformation is completely arrested and the endometrium remains in a state of "arrested proliferation" (Bayer 1965). On continued medication this state gives way to progressive glandular atrophy and simultaneous decidualization of stroma, until after about 3 months the picture we see is one of "arrested secretion" (Fig. 1) (Winter and Pots 1956; Scommegna et al. 1970). At this stage there is as good as no secretion of glycoproteins at all, and the activity of almost all of the enzymes present in the endometrium is down to a very low level, particularly that of β-glucuronidase. The spiral arteries are very poorly developed, as are the cell organelles in nucleus and cytoplasm when viewed under the elec-

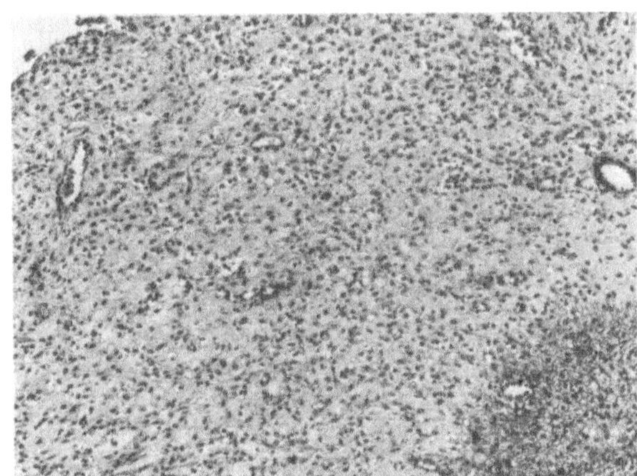

Fig. 1. Arrested secretion of corpus endometrium with decidualization of stroma and glandular atrophy following 3-month administration of a gestagen compound. H and E stain. x 93

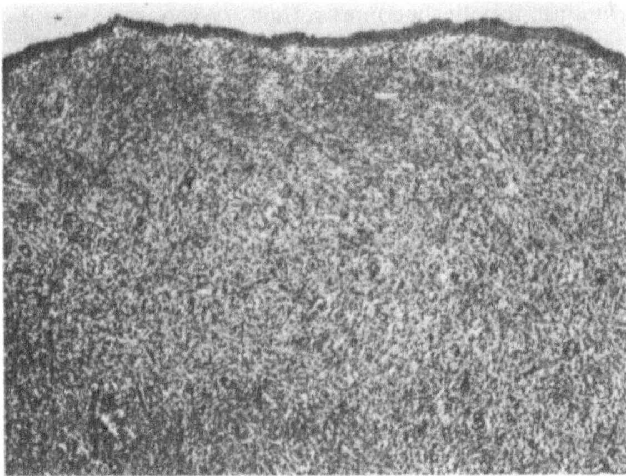

Fig. 2. Fibrous atrophy of corpus endometrium with complete glandular atrophy following 9-month administration of gestagen. H and E stain. x 39

tron microscope. The decidual transformation is more obvious after 19-nor-testosterone derivatives than after progesterone derivatives. If administration is continued, this state of suspended or arrested secretion develops into a true, irreversible atrophy accompanied by hyalinization of the stroma (Fig. 2). Since even under physiologic conditions the glandular epithelium reacts earlier and more sensitively than the stroma to progesterone, the epithelium probably becomes refractory first when progesterone levels become pathologic, with the stroma following after a period of decidualization. Tissue fibrinolysis measurements have revealed that plasmin activity is blocked at this stage (Staffeldt et al. 1969).

Recent clinical, morphological observations with depot gestagens or the minipill have shown that they may produce similar, high-degree atrophy of the endometrium and complete glandular atrophy from the 4th month of treatment on, and sometimes even earlier though seldom later. Even after the drug is discontinued these atrophies may remain irreversible for long periods of time. Analogously, even high-degree adenomatous hyperplasias following months' long treatment with depot gestagens can go, via secretory transformation (Fig. 3) to irreversible atrophy (which in this case is a therapeutic tool. Kistner 1959). Even adenocarcinomas are transformed by high doses of gestagens, will sometimes regress, and show a clear decrease in DNA synthesis (Kohorn 1976; Gerulath and Borth 1977; Seewald et al. 1977).

The unfavorable general effects observed in young females when systemic gestagens were used led to their *local application,* by means of an intrauterine device, a T-shaped pessary with a depot in the leg of the T which gives off 65 μg progesterone daily into

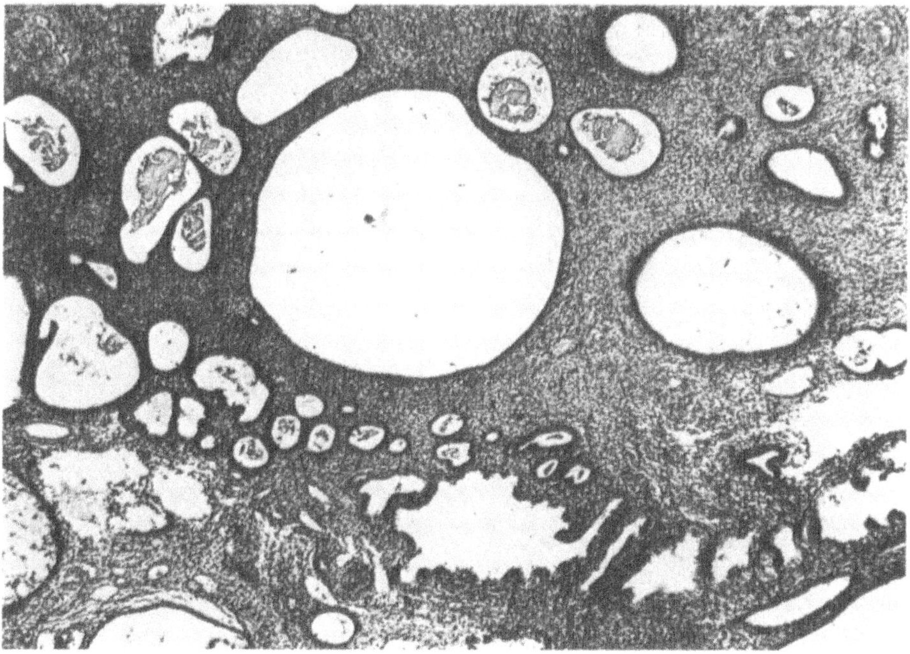

Fig. 3. Secretory transformation of adenomatous cystic hyperplasia of corpus endometrium following several months' therapy with depot gestagens. H and e stain. x 44

the uterine cavity. This device, in addition to its known paracrine effects, causes a perifocal decidualization with glandular atrophy (arrested secretion) in a limited area which is very sharply defined, below which we see normally proliferating and secreting functionalis (Dallenbach-Hellweg and Sievers 1975) (Fig. 4). Thus with intrauterine application we have succeded in limiting gestagen effects to a proscribed target area. The perifocal, arrested secretion seen differs on the one hand from the complete suspension of endometrial secretion caused by oral or parenteral gestagen administration, and on the other from the mechanical decidualization caused by nonmedicated intrauterine devices. The contraceptive effect is just as reliable as that given by generalized oral administration, with a Pearl index of under 1%, since the implantation of a blastocyst depends on the condition of the upper endometrium layers. And the positive difference seen with this device is that the basal layers of the endometrium, which are crucial to regeneration, remain unharmed. Reliability of contraception with this device is higher than that provided by normal pessaries because of the glandular atrophy it leads to, and the dangers of generalized gestagen effect are avoided.

Synthetic gestagens are also known to lead to atypical adenomatous hyperplasias of the *endocervix*, which are of a far greater extent than those seen in pregnancy, in both quantitative and qualitative terms (Fig. 5), and give a radically different picture to the endometrial atrophy seen after pure gestagen therapy. The effects of synthetic gestagens can be seen macroscopically in extirpated uteri; the cervical canal is usually wide and is filled with thick, viscous mucus (Lebech et al. 1969; Moghissi and Marks 1971). This mucus is reduced and biochemically altered (Gregoire and Ustay 1969;

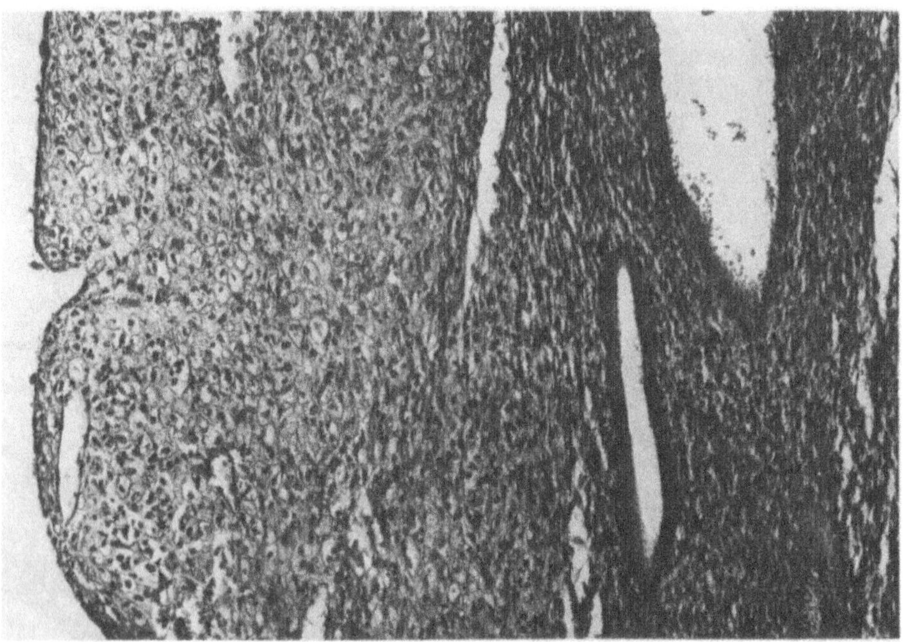

Fig. 4. Endometrium following use of intrauterine device containing gestagen. Note arrested secretion in the upper layer, and sharply defined border to normally proliferating endometrium. H and E stain. x 148

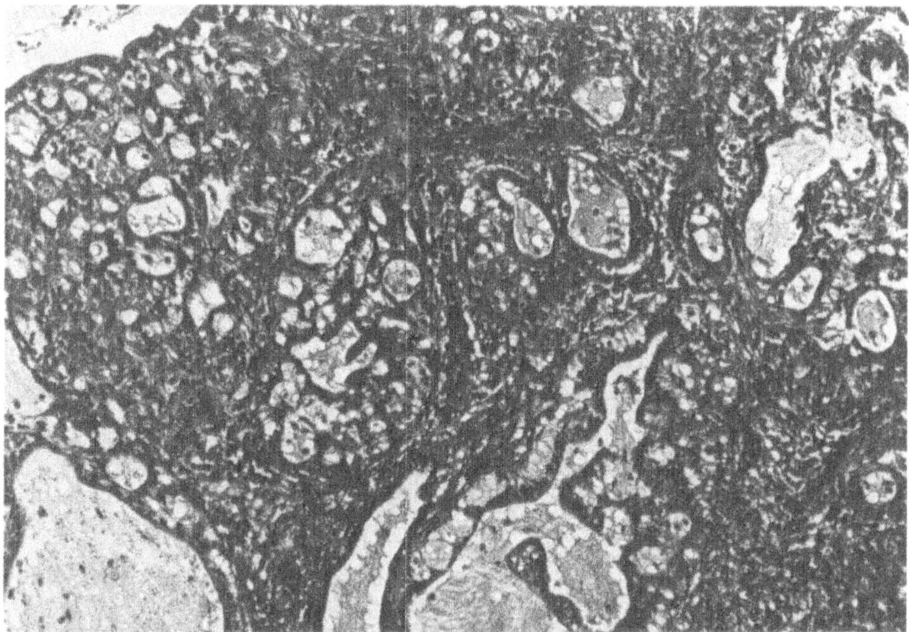

Fig. 5. Adenomatous hyperplasia of endocervix after the use of gestagen-containing contraceptive. Note small alveolar branching and surrounding reserve cell hyperplasia; stroma is almost totally atrophied. H and E stain. x 146

Elstein 1970). What seems histologically to be a hypersecretion, the glands full to bursting, is really only a result of this thickening process and the retention of the secretion in the glands. The higher the gestagen dosage and/or potency, the more marked are these changes (Maqueo et al. 1966). The cervical glandular epithelium sometimes evinces multiple rows, or signs of reserve cell hyperplasia; areas of small alveolar branching are also seen (Candy and Abell 1968).

An increasing number of pseudosarcomatous proliferations in the *myometrium* have also been observed following treatment with synthetic gestagens, e.g., leiomyomas that are particularly rich in cells, quick growing or atypical (Fechner 1968), or endolymphatic stromal myosis (Behary et al. 1976).

References

Bayer R (1965) Die Endometriumreaktion auf Chlormadinoazetat bei Frauen mit monophasischer und biphasischer Zyklussteuerung. I. Langzeitbehandlung mit Kleindosen. II. Langzeitbehandlung mit ansteigenden Dosen zwischen 10 mg und 30 mg Chlormadinonazetat. Z Geburtshilfe Gynaekol 164:47–62

Behary CM, Dollberg L, Czernobilsky B (1976) Endolymphatic stromal myosis in two patients on progestagen therapy. Contraception 13:1

Candy AMR, Abell MI (1968) Progestogen-induced adenomatous hyperplasia of the uterine cervix. J Am Med Assoc 203:323

Dallenbach-Hellweg G, Sievers S (1975) Die histologische Reaktion des Endometrium auf lokal applizierte Gestagene. Virchows Arch [Pathol Anat] 368:289–298

Elstein M (1970) The proteins of cervical mucus and the influence of progestagens. J Obstet Gynaecol Br Commonw 77:443

Fechner RE (1968) Atypical leiomyomas and synthetic progestin therapy. Am J Clin Pathol 49:697

Gerulath AH, Borth R (1977) Effect of progesterone and estradiol-17-β on nucleic acid synthesis in vitro in carcinoma of the endometrium. Am J Obstet Gynecol 128: 772

Gregoire AT, Ustay K (1969) Effect of chlormadinone on amount of human cervical mucus and its glycogen content. Fertil Steril 20:938

Kistner RW (1959) Histological effects of progestins on hyperplasia and carcinoma in situ of the endometrium. Cancer (Philad.) 18:1563

Kohorn EI (1976) Gestagens and endometrial carcinoma. Gynecol Oncol 4:398

Lebech PE, Svendsen PA, Ostergaard E, Koch P (1969) The effects of small doses of megestrol acetete on the cervical mucus. Acta Obstet Gynecol 48:22

Maqueo M, Azuela JC, Calderon JJ, Goldzieher JW (1966) Morphology of the cervix in women treated with synthetic progestins. Am J Obstet Gynecol 96:994

Moghissi KS, Marks C (1971) Effects of microdose norgestrel on endogenous gonadotropic and steroid hormones, cervical mucus properties, vaginal cytology, and endometrium. Fertil Steril 22:424

Scommegna A, Pandya GN, Christ M, Lee AW, Cohen MR (1970) Intrauterine administration of progesterone by a slow releasing device. Fertil Steril 21:201

Seewald H-J, Böhm W, Przesang R (1977) Ergebnisse der hochdosierten Gestagentheraphie bei Carcinoma corporis uteri. Zentralbl Gynaekol 99:392

Staffeldt K, Kreuzer G, Hoffbauer H (1969) Histologische und gerinnungsanalytische Untersuchungen zur Wirkung der niedrig dosierten Gestagen-Dauerapplikation auf das Endometrium. Arch Gynaekol 207:50

Winter G, Pots P (1956) Morphologische Untersuchungen über die medikamentöse Transformation des Endometriums. Z Geburtshilfe Gynaekol 147:44

Morphological Effects of Exogenous Gestagens on Abnormal Human Endometrium

A. Ferenczy

Abstract

Progestagens given in appropriate doses and for appropriate time may control endometrial hyperplasia and a certain proportion of carcinoma. Progestagen-induced secretory differentiation is followed by a decrease in DNA-synthesis, arrest in mitosis, prolongation of the tissue turnover time, and an increase in autophagy and cellular and tissue necrosis. Secretory conversion, inhibition of DNA synthesis, and mitosis are features associated with an intact receptor-hormone binding mechanism. Autophagy is a consequence of gestagen-induced secretory activity of epithelial cells, whereas cellular and coagulative necrosis are a reflection of the nonspecific cytotoxic effect of pharmacologic doses of progestagens. The presently available data suggest that hyperplastic and neoplastic endometria are made of a heterogeneous cellular population in which receptor-containing hormone-sensitive cells are admixed with receptor-free, and consequently hormone-insensitive cells. The latter are independent of hormonal influence, grow autonomously and their number presumably increases paralleling the severity of endometrial hyperplasia and carcinoma. They are responsible for recurrences and/or the nonresponsiveness of most adenocarcinomas and atypical hyperplasias to gestagen therapy.

Progestational therapy of hyperplastic and neoplastic endometria is based on our knowledge of the close interaction between the morphobiochemical-enzymatic changes in the normal endometrium and the ovarian hormones of the menstrual cycle. Progesterone induces secretory conversion of estrogen-primed, normal proliferative gland-lining cells. This involves the ergastoplasm–Golgi–mitochondria–glycogen complex and results in the production of glycoprotein-rich secretions, inhibition of DNA synthesis and mitotic activity, and predecidual transformation of stromal fibroblasts (Ferenczy and Richart 1974; Wynn 1977).

It has been proposed that most endometrial hyperplasias (EH) and at least a certain proportion of carcinomas may retain the steroidobiochemical properties of normal endometrium, and consequently may respond to progestagenic therapy (Kelly and Baker 1961). Large doses of progestagens such as medroxyprogesterone acetate (Provera), hydroxyprogesterone caproate, megestrol acetate, have been shown to control EH and about one-third of adenocarcinomas clinically (Kohorn 1976) and experimentally (Nordqvist 1969; Hertz 1974), both in vitro and in vivo. Changes have been demonstrated by means of histology (Kistner 1959; Anderson 1972; Hustin 1973; Rosier and Underwood 1974), histochemistry (Hustin 1973; Thiery and Willighagen 1967), electron microscopy (Richart and Ferenczy 1974; Ferenczy 1977; Sekiya et al. 1977), liquid scintillation radioautography (Nordqvist 1969; Gerolath and Borth 1977), in vitro historadioautography, (Siracky et al. 1978) and steroid hormone receptor analyses (Vihko et al. 1978).

In vivo and in vitro morphobiochemical studies correlated the cellular response of EH and carcinoma to progestagens with the uptake and specific binding of estradiol-17β (E_2) (Vihko et al. 1978) and progesterone (P) (Vihko et al. 1978). In EH, receptors for both E_2 and P are uniformly present and their levels are closely similar to those found in cyclic, proliferative endometrium (Vihko et al. 1978; Tseng et al. 1977; Shyamala and Ferenczy, to be published) (Fig. 1). In most adenocarcinomas, sex-steroid receptors are demonstrated, although at lower concentrations than either in proliferative or hyperplastic endometrium (Fig. 1). Also, there is a trend toward higher concentrations in well-differentiated, and lower concentrations in poorly differentiated carcinomas (Vihko et al. 1978). These observations are in agreement with the clinical findings that hyperplastic endometrium and well-differentiated carcinomas are more likely to respond to progestagen therapy than are poorly differentiated neoplasms (Kohorn 1976; Kistner 1959; Anderson 1972; Hustin 1973; Rosier and Underwood 1974).

A correlation has been demonstrated between objective tissue regression and the duration of the therapy. The minimum effective treatment period is at least 12 weeks (Kohorn 1976; Reifenstein 1974). The rate of objective response of endometrial carcinomas is similar (about 30%) for various progestagens, providing that a high initial loading dose (400–1000 mg/week depending on the compound used) is given in the first few weeks (Kohorn 1976). Also, the secretory conversion, decrease in mitosis and tissue regression are of the same nature and magnitude whether or not the progestagen is used parenterally or intracavitarily (Kohorn 1976; Hustin 1973).

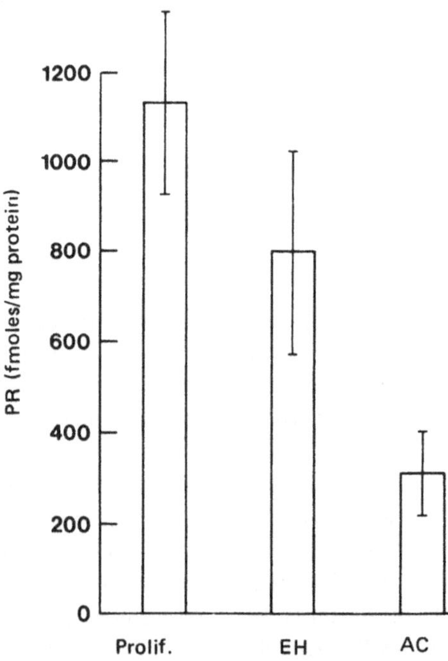

Fig. 1. Progesterone receptors (PR) are elevated in cyclic, proliferative endometrium (Prolif.; 9 cases) and hyperplasia (EH; 5 cases), whereas in carcinomas (AC; 18 cases) PR concentrations, although present in 95 % of the cases are considerably lower. Progesterone receptors were measured by the dextran-coated charcoal binding assays using tritiated R5020 ligand

The dose-dependent sensitivity of EH and carcinoma to progestagens in in vitro situation (Nordqvist 1969; Sekiya et al. 1977; Gerolath and Borth 1977) provides evidence that these compounds act directly at a cellular level. The data suggest, furthermore, that gestagen-induced changes occur independently of immunosuppressive mechanisms or endogenous hormonal environment. Indeed, gestagen-treated carcinoma and hyperplasia patients have normal or nonspecific variations of immunoelectrophoretic profiles, peripheral blood lymphocyte count and endogenous pituitary and steroid hormone levels (Anderson 1972). Moreover, the changes observed in progestagen-treated abnormal endometria reflect an intact progesterone receptor mechanism (Ferenczy et al. 1978), since similar alterations are induced by P in normal, postovulatory endometrium (Ferenczy and Richart 1974; Wynn 1977).

This contention seems to be supported, furthermore, by a significant reduction in E_2 R and PR levels and an increase in 17β-hydroxysteroid dehydrogenase (E_2 DH) in all cases of EH and in about 50% of endometrial adenocarcinomas treated with Provera (Vihko et al. 1978; Tseng et al. 1977). Both E_2R and E_2DH activities are regulated by P, via an intact receptor mechanism. A decrease in PR concentrations is a reflection of the inhibitory effect of Provera on E_2-stimulated DNA-dependent RNA synthesis. Indeed, receptors for both E_2 and P are produced by E_2 (Clark et al. 1977).

In our laboratory we have studied the ultrastructural changes and proliferation kinetics of gestagen-treated EH and carcinoma in 49 women (Table 1). The processing, preparation and interpretation of the material for electron microscopy and in vitro historadioautography are described in detail in previous publications (Ferenczy and Richart 1974; Wynn 1977; Ferenczy et al. 1979). In all cases the original diagnosis was ascertained by endometrial biopsy and the patients were followed by endometrial biopsy at 1 to 3-week intervals for 12 weeks. In women with a diagnosis of carcinoma, endocervical curettings were performed to rule out cervical involvement. In no cases the endocervical curettings were positive for carcinoma.

As shown in Table 1, most endometrial hyperplasias including those with severe cytologic atypia demonstrated significant gestagen-related morphological response. In

Table 1. Progestagen therapy of endometrial hyperplasia and carcinoma[a]

No patients	Age (yrs)	Diagnosis	Provera 100 mg/week x 12 weeks		Depo-Provera 1 gm/week x 12 weeks	
			Response	No response	Response	No response
12	54 ± 4	CGH–AH	12	0		
10	57 ± 2	AAH	7	3		
27	63 ± 2	Carcinoma				
		Grade 1			10	2
		Grade 2			6	5
		Grade 3			1	3
Total 49						

[a] Provera: medroxyprogesterone acetate; CGH: cystic glandular hyperplasia; AH: adenomatous hyperplasia; AAH: atypical adenomatous hyperplasia; Grade 1: well-differentiated; Grade 2: moderately differentiated; Grade 3: poorly differentiated carcinoma.

contrast, only 17 of 27 cases of carcinoma showed morphological response to therapy; the rate of response was correlated with the degree of differentiation. The 10 patients who demonstrated no tissue response to Provera by the 4th week of therapy were excluded from the study and received radiotherapy. Progestagen therapy of the carcinoma patients was chosen intially because surgery was either refused by the patient (4 cases) or contraindicated because of poor cardiorespiratory functions (11 cases). The remainder of 12 patients agreed to be treated with gestagens for a period of 12 weeks. In this group, after termination of the study, a bilateral salpingo-oophorectomy and hysterectomy were performed. External radiotherapy was given if a greater than the inner one-third of the myometrium was invaded by the tumor. Patients in whom surgery could not be performed received radiotherapy after termination of the 12 weeks gestagen therapy.

The morphological response of hyperplastic and neoplastic endometria to progestagen therapy include: (1) secretory conversion, (2) growth supression, and (3) degeneration (Ferenczy et al. 1978). These alterations lead to tissue regression (Fig. 2). The observations in the human are in agreement with proliferation kinetic studies in experimental animals (Epifanova 1971). In the animal model, secretory differentiation of the endometrium is associated with arrest of cell growth, followed by death of tissue (Epifanova 1971).

As early as 10–14 days after administration of large doses of Provera, secretory conversion of the cytoplasmic substance of hyperplastic and neoplastic gland lining cells is observed. This is evidenced by the marked development of the ergastoplasm–Golgi–mitochondria–glycogen complex (Richart and Ferenczy 1974; Ferenczy 1977; Sekiya et al. 1977) (Fig. 3a and b). Intracellular glycoprotein synthesis is followed by apocrine secretory activity whereby large portions of the apical cytoplasm are detached and expulsed into the glandular lumen (Fig. 3c). Such secretory phenomenon results in a considerable reduction in the height of epithelial cells. Secretory differentiation is accompanied by a dramatic decrease and eventual arrest of mitotic activity with a sharp drop of the mitotic index (Hustin 1973). Mitotic arrest parallels a profound decrease in nuclear DNA and RNA synthesis as demonstrated by in vitro liquid scintillation radioautography (Nordqvist 1969).

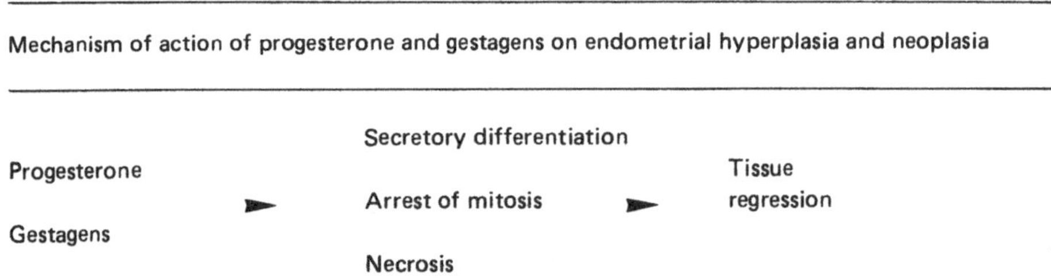

Fig. 2. The predominant events responsible for gestagen or progesterone-induced endometrial tissue regression are the arrest of mitosis followed by secretory differentiation and glandular "exhaustion". Necrosis is a less important contributor to tissue regression under in vivo conditions

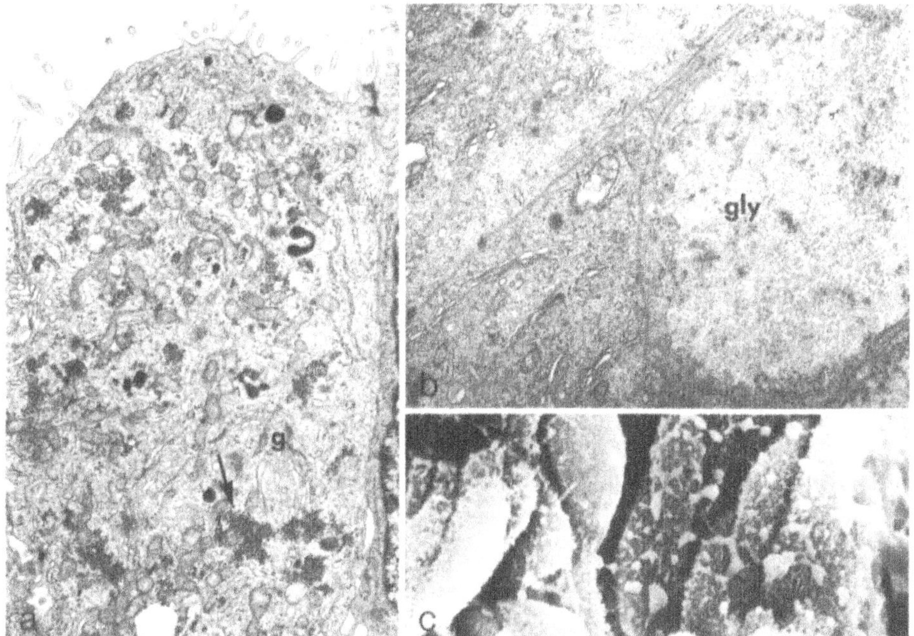

Fig. 3. Electron microscopy of well-differentiated adenocarcinoma, invasive of the endometrial mucosa after 4 weeks of Depo-Provera treatment. **a** Glycogen granules *(arrow)* intermingle with hypertrophic Golgi *(g)* and mitochondria. **b** Atypical adenomatous hyperplasia fater 4 weeks of Provera treatment. Note massive accumulation of electronpalc glycogen granules *(gly)* and hypertrophied Golgi. **c** Scanning electron microscopy of cystic glandular hyperplasia after 8 weeks of Provera treatment. Secretory activity is evidenced by numerous secretory droplets over cell surfaces. Ciliated cells are absent and surface microvilli are shorter and fewer in number than in untreated hyperplasia. x 1208

Using short term incubation in vitro historadioautography with double labeling with radiothymidine, a nucleoprotein precursor we have observed a two to three fold decrease in labeling index (the ratio of the number of labeled cells to all unlabeled cells determined from a sample of 2,000 cells) in progestagen-treated neoplastic endometrial tissue (Fig. 4). Due to the decreased number of cells engaged in active DNA-synthesis and the unchanged duration of the DNA-synthesis phase, the potential doubling time or tissue turnover rate of gestagen-treated endometria is considerably lengthened (Fig. 4). Similar observations were made by others (Siracky et al. 1978). In experimental animals, progestagens presumably induce the epithelial cells to enter into the G_0, non dividing stage of the cell cycle (Zhinkin and Samoshkina 1967). It is conceivable that a similar phenomenon operates in the human. Indeed, decreased DNA synthesis coincides with the conversion of tetraploid (Hustin 1976) and near-diploid, aneuploid (Wagner et al. 1967) nuclear content to diploid DNA mode of normal endometrial cells (Hustin 1976). In the hyperplastic endometrium, estrogen-related ciliogenesis is arrested and surface microvillogenesis is decreased (Richart and Ferenczy 1974; Ferenczy 1977). Further tissue involution occurs later (4 to 12 weeks of treatment)

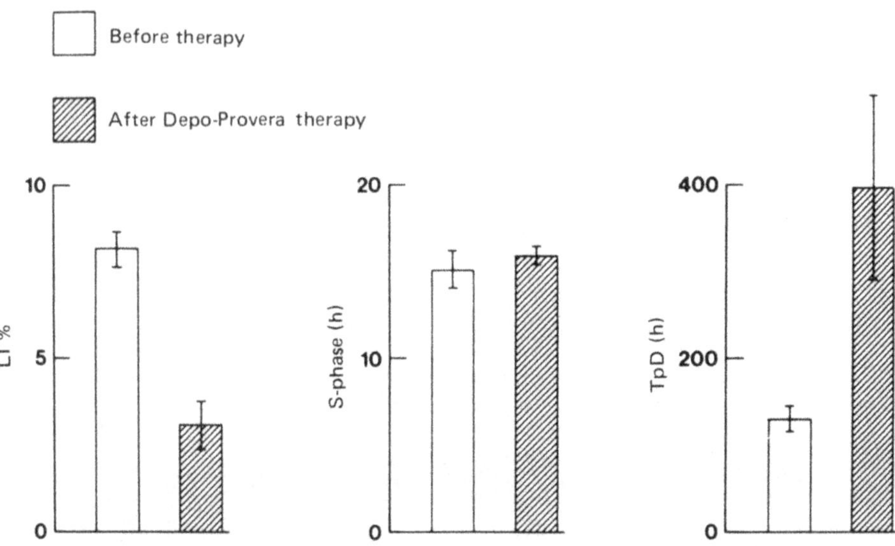

Fig. 4. Mean proliferation kinetics of endometrial carcinoma (17 cases) before and after 12 weeks of Depo-Provera therapy. Due to a threefold decrease in LI (labeling index) and the unchanged duration of DNA-synthesis phase (S-phase), the potential doubling time of neoplastic tissue (TpD) is considerably lengthened ($TpD = \lambda\,(0.7)\,\frac{S}{LI}$)

by autophagy, single cell necrosis (apoptosis) and coagulative necrosis (Richart and Ferenczy 1974; Ferenczy 1977).

Autophagic activity may be stimulated by a wide variety of physiologic and pathologic conditions. In gestagen-treated abnormal endometria autophagy is considered to be a consequence of secretory conversion of epithelial cells (Ferenczy et al. 1978). Autophagic bodies are membrane-bound structures in which are incorporated intracytoplasmic organelles and various inclusions (Fig. 5a and b). Autophagic bodies fuse with primary lysosomes, the membranes of which are presumably labilized by the membrane-toxic effect of pharmacologic doses of progestagens (Weissmann 1964). Membrane permeability facilitates the release of lysosomal enzymes such as acid phosphatase into the enzyme-free autophagic bodies. Upon the destructive action of the acid hydrolases, the incorporated cytoplasmic elements are digested producing empty vacuoles (Fig. 5b). As a result of this process, portions of the cytoplasmic substance are removed by lysosomal autodigestion.

It has been suggested that lytic enzymes may also diffuse into the intra- and intercellular spaces of glandular and stromal cells as well as the vascular endothelium resulting in cellular dehiscence (Ferenczy et al. 1978). Although neoplastic or hyperplastic endometria have not yet been studied by means of ultracytochemistry using acid phosphatase, the available ultrastructural observations (Richart and Ferenczy 1974; Ferenczy 1977; Sekiya et al. 1977) suggest that the proposed morphobiochemical mechanisms are likely to operate in progestagen-treated abnormal endometria. Indeed, acid phosphatase content is elevated in progestagen-treated as compared to nontreated endometrial adenocarcinoma in the human (Thiery and Willighagen

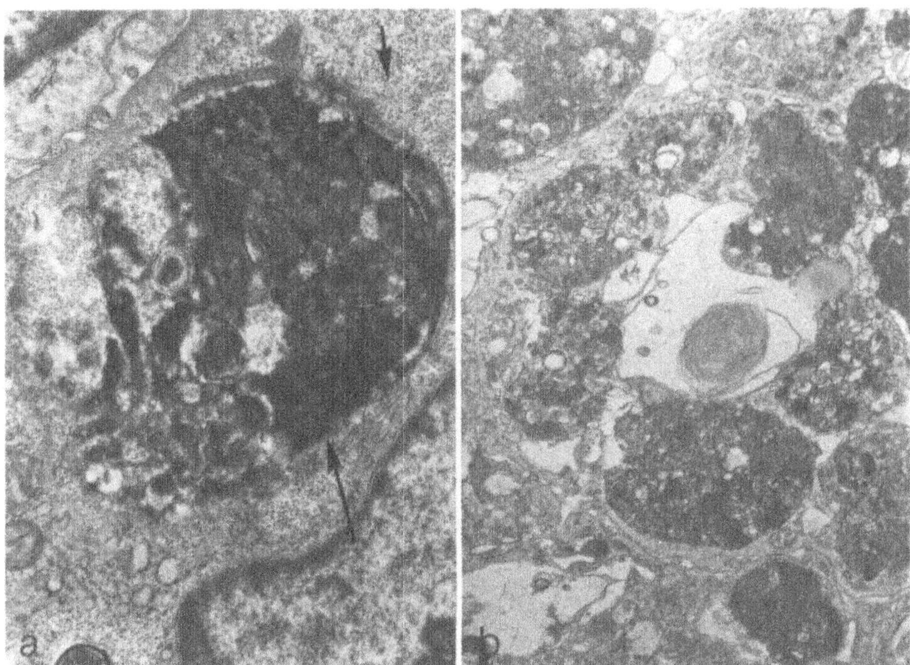

Fig. 5. Moderately differentiated adenocarcinoma after 12 weeks of Depo-Provera therapy. **a** Detailed view of a giant membrane-bound autophagosome in which partly digested amorphus material *(long arrow)* intermingle with relatively intact organelles and membranes. Glycogen granules *(short arrow)* are abundant in the surrounding cytoplasm. **b** Atypical adenomatous hyperplasia after 5 weeks of Provera therapy. Intracytophagolysosomalvesicles and membranes demonstrate varying degree of enzymatic digestion. In the center is an empty vacuole representing the end stage of enzymatic autodigestion. x 5520

1967) and progesterone treatment of cultured rat uterine adenocarcinoma cells lead to prominent autophagic activity (Sekiya et al. 1977).

Increased autophagic activity is associated with increasing numbers of individual epithelial cells undergoing total cell necrosis or apoptosis (Richart and Ferenczy 1974; Ferenczy 1977; Ferenczy et al. 1978). The basic cellular changes in apoptotic cells are consistent with dying cells. These include karyorrhexis, karyolysis with cytoplasmic and organellar shrinkage. Many of these degenerated cells are eventually removed by phagocytosis by neighboring neoplastic cells (Fig. 6a) as well as by macrophages, and digested by hydrolytic lysosomal enzymes (Richart and Ferenczy 1974; Ferenczy 1977) (Fig. 6b).

Individual cell necrosis is not confined to progestagen-treated abnormal endometria but is seen also in untreated endometrial neoplasias as well as in extragenital cancers (Searle et al. 1975). Apoptosis is regarded as a biologic process regulating cellular growth by providing indirect control of mitosis by continuous cell loss. In other words, as new cells are produced by mitotic division, others undergo necrosis. The comparative abundance of apoptotic cells in endometrial lesions treated with pro-

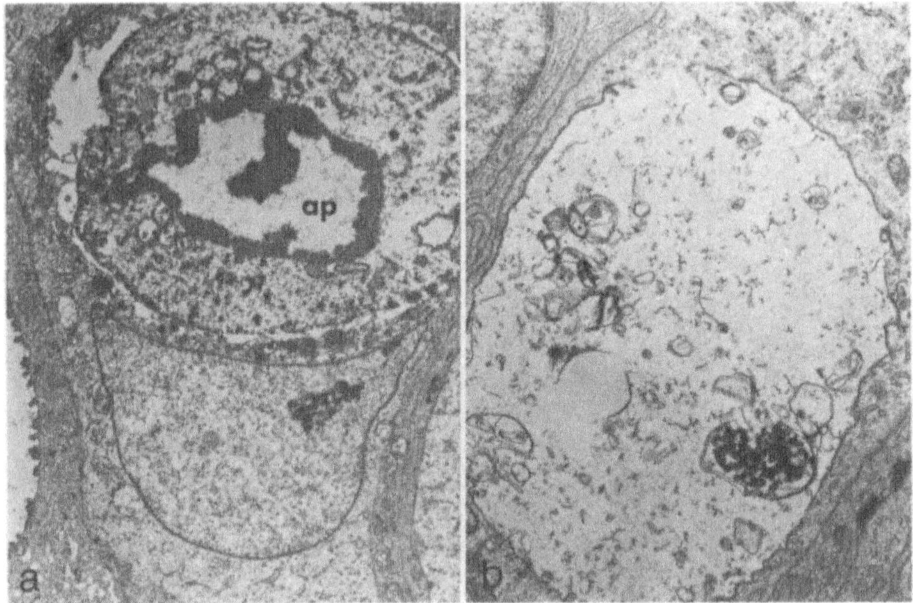

Fig. 6. Moderately differentiated adenocarcinoma after 12 weeks of Depo-Provera therapy. **a** An apoptotic cell *(ap)* with degenerated nucleus and cytoplasm has been phagocytosed by a neighboring, intact neoplastic cell x 5175. **b** Following enzymatic destruction of a degenerated, apoptotic body a giant intracellular, virtually empty membrane-bound vacuole is produced. x 4830

gestagens, however, suggests that the accelerated cell-death rate may contribute to the regression of neoplastic and hyperplastic epithelia (Ferenczy et al. 1978).

Coagulation necrosis in progestagen-treated hyperplasia and in particular carcinoma are thought to be related to hormone therapy rather than to the initial biopsy or manipulation in delivering progestagens into the endometrial cavity (Ferenczy et al. 1978). Whether progestagen-induced cytotoxicity in vivo is a nonspecific or receptor-mediated effect remains to be determined, but the former is more likely. In vitro studies using short-term incubation systems suggest that progesterone at high concentrations produces tissue necrosis, whereas at low concentrations, tissue differentiation (Nordqvist 1969; Gerolath and Borth 1977). Although in vitro culture systems do not necessarily reflect the in vivo process, the maximal inhibition of DNA synthesis at a concentration of 80 μg/ml is comparable to the plasma concentration of progesterone (80 μg/ml) in patients with carcinoma who receive 150–200 mg of progesterone daily (Gerolath and Borth 1977). This dosage regimen seems to provide the maximal clinical and morphological response (Gerolath and Borth 1977). It is interesting to note that the in vitro, nonspecific necrotizing effect of progestagens may also be reproduced by estradiol alone at high concentrations (20 μg/ml–40 μg/ml) (Nordqvist 1969; Gerolath and Borth 1977).

Under the influence of large doses of progestagens given for more than 5 weeks, the spindle-shaped endometrial stromal cells of endometrial hyperplasia and intramucosal adenocarcinoma may acquire ultrastructural characteristics similar (Ferenczy

1977) to those observed in normal premenstrual, predecidual cells or in gestational decidual cells (Ferenczy and Richart 1974; Wynn 1977). These include nuclear enlargement, hyperplasia of the granular endoplasmic reticulum and Golgi systems and accumulation of glycogen. The cells are engaged in sialic acid-rich glycoprotein synthesis. Based on experimental studies on the prepuberal rat uterus (Sananes et al. 1976) in which stromal decidualization occurs by progesterone-mediated prostaglandin $F_{2\alpha}$ stimulation, it is possible that progestagens may induce similar biologic sequences of events in the stromal cells of endometrial hyperplasia and early neoplasia. The premenstrual, decidual cells have been shown to phagocytose and degrade extracellular collagen within acid phosphatase-containing cytophagolysosomes (Henzl et al. 1972). Lysosomal digestion of preexisting extracellular collagen fibers, followed by enzymatic autodestruction of the decidualized cells themselves may contribute to the regression of progestagen-treated hyperplastic and neoplastic endometrial stroma (Richart and Ferenczy 1974; Ferenczy 1977).

Most poorly differentiated neoplasms and lesions from older patients associated with an atrophic endometrium, fail to demonstrate morphological and clinical response to progestagen therapy (Kohorn 1976; Anderson 1972; Hustin 1973; Kistner et al. 1965). In these lesions, nuclear aneuploidy persists (Hustin 1976) secretory differentiation is absent or abortive and progesterone receptors are either absent or inconspicuous (Vihko et al. 1978; Shyamala and Ferenczy, to be published). Similar lack of progestagen-related morphological changes, although with a focal rather than diffuse distribution are seen in 50% to 70% of the better differentiated carcinomas as well as in 10% to 50% of adenomatous hyperplasia with severe cytologic atypia. These findings suggest that advanced hyperplasia and neoplastic endometria with varying degrees of cytologic differentiation are composed of a heterogeneous cell population, in which steroid-sensitive cells intermingle with steroid-insensitive cells. The latter are presumably independent of hormonal influence, grow autonomously and are responsible for the high rate of recurrence and/or lack of response of endometrial carcinomas treated by progestagen therapy.

Acknowledgements

The author is grateful to Miss Rosemary De Marco for typing the manuscript.

This work was supported in part by grant MA-3157 from the Medical Research Council of Canada.

References

Anderson DG (1972) The possible mechanisms of action of progestins on endometria adenocarcinoma. Am J Obstet Gynecol 113:195

Clark JH, Hseuh AJW, Peck EJJr (1977) Regulation of estrogen receptor replenishment by progesterone. Ann NY Acad Sci 286:161

Epifanova OI (1971) In: Barerga R (ed) The cell cycle and cancer Dekker, New York, p 145 Effects of hormone on the cell cycle.

Ferenczy A (1977) Ultrastructural pathology of the uterus. In: Wynn RM (ed) Biology of the uterus. Plenum Press, New York, p 545

Ferenczy A, Richart RM (1974) Female reproductive system: Dynamics of scan and transmission electron microscopy. Wiley, New York

Ferenczy A, Guralnick M, Gelfand MM (1978) Observed changes in endometrial carcinoma during hormone induced regression. Rev Endocrine Related Cancer 1:21
Ferenczy A, Bertrand G, Gelfand MM (1979) Proliferation kinetics of human endometrium during the normal menstrual cycle. Am J Obstet Gynecol 133:859
Gerlath AH, Borth R (1977) Effect of progesterone and estradiol-17β on nucleic acid synthesis in vitro in carcinoma of the endometrium. Am J Obstet Gynecol 128: 772
Henzl MR, Smith RE, Boost G, Tyler ET (1972) Lysosomal concept of menstrual bleeding in humans. J Clin Endocrinol Metab 34:860
Hertz R (1974) Pharmacology of the endometrium. Gynecol Oncol 2:264
Hustin J (1973) Hormonal therapy of endometrial cancer: Effects of large doses given by parental or intracavitary routes. In: Brush MG, Taylor RW, Williams DC (eds) Symposium on endometrial cancer. London, p 246
Hustin J (1976) Morphology and DNA content of endometrial cancer nuclei under progestogen treatment. Acta Cytol (Baltimore) 20:556
Kelly RM, Baker WH (1965) The role of progesterone in endometrial cancer. Cancer Res., 25:1190
Kistner RW (1959) Histologic effects of progestins on hyperplasia and carcinoma in situ of the endometrium. Cancer 12:1106
Kistner RW, Griffith CT, Crain JM (1965) Use of progestational agents in the managment of endometrial cancer. Cancer 18:1563
Kohorn EI (1976) Gestagens and endometrial carcinoma. Gynecol Oncol 4:398
Nordqvist SRB (1969) Hormonal responsiveness of human endometrial carcinoma studied in vitro and in vivo. Thesis, Lund, Sweden
Reifenstein EC Jr (1974) The treatment of advanced endometrial cancer with hydroxyprogesterone caproate. Gynecol Oncol 2:377
Richart RM, Ferenczy A (1974) Endometrial morphologic response to hormonal environment. Gynecol Oncol 2:180
Rosier JG, Underwood PB (1974) Use of progestational agents in endometrial adenocarcinoma. Obstet Gynecol 44:60
Sananes N, Baulieu EE, Le Goascogne C (1976) Proslaglandin(s) as inductive factor of decidualization in the rat uterus. Mol Cell Endocrinol 6:153
Searle J, Lawson TA, Abbott PJ, Harmon B, Keer JFR (1975) The spontaneous occurence of apoptosis in squamous careinomas of the uterine cervix. J Pathol 116:129
Sekiya S, Takeda B, Kikuchi Y, Sakaguchi S, Takamizawa H (1977) Morphologic and enzyme-cytochemical changes in uterine adenocarcinoma cells of a rat by direct application of progesterone in vitro. Gynecol Oncol 5:5
Shyamala G, Ferenczy A (to be published) Steroid receptors in normal and abnormal endometria. Cancer Res.
Siracky J, Schveiner P, Siracka E, Matoska J (1978) Cell proliferation kinetics and nuclear morphology in endometrial cancer under progesterone treatment. Neoplasma 25:5
Thiery M, Willighagen RGJ (1967) Enzyme histochemistry of adenocarcinoma of the endometrium including hormone-induced changes. Am J Obstet Gynecol 99:173
Tseng L, Gusberg SB, Gurpide E (1977) Estradiol receptor and 17β-dehydrogenase in normal and abnormal human endometrium. Ann NY Acad Sci 286:190
Vihko R, Pollow K, Robel P (1978) Receptors for steroids in human endometrium. (eds) UICC (Union Int Contre Cancer) Tech Rep Ser 42:63–112
Wagner D, Richart RM, Turner JY (1967) Deoxyribonucleic acid content of presumed precursors of endometrial carcinoma. Cancer 20:2067
Weissmann G (1964) Labilization and stabilization of lysosomes. Fed Proc 23:1038
Wynn RM (1977) Histology and ultrastructure of the human endometrium. In: Wynn RM (ed) Biology of the uterus. Plenum Press, New York, p 341
Zhinkin LN, Samoshkina NA (1967) DNA synthesis and cell proliferation during the formation of deciduomata in mice. J Embryol Exp Morphol 17:598

Differential Action of Sex Steroids on Ovary and Adrenal Cortex of Syrian Golden Hamster

Gertraud Zieger, W. Zieger and B. Kubatsch

Abstract

Materials and Methods

The methods used were gravimetry, light microscopy and statistical analysis of the values obtained.

Each of the following hormones -- medroxyprogesterone acetate (MP), cyproterone acetate (CA), dydrogesterone (DY), or testosterone propionate (TP) – was injected in increasing dosage into groups of 16 female hamsters starting on the 4th day after their birth and continuing to their 8th week of age. Weekly doses per animal: MP, 0.1–0.8 mg, CA, 1.25–10 mg, DY, 1.25–10 mg, TP, 0.25–2 mg; 16 control animals received only dissolving agent. Immediately after treatment ceased, one-half of each group was killed; the remaining killed after 4 weeks. We weighed the animals, their ovaries, and adrenals, and fixed these endocrine organs in Bouin's solution.

Results

After 8 weeks of treatment the ovaries of the TP-treated animals weighed significantly less than those of the controls. Histologically, primary and secondary follicles were occasionally present; tertiary follicles or corpora lutea could not be found. Similar changes were evident in most of the MP-treated animals.

When we compared the ovaries of 8-week-old animals with those from the 12-week-old animals of the same group, we could detect gravimetrically and histologically differences in the animals treated with MP and TP. In both groups, the ovaries showed an impressive recovery during the 4-week pause in treatment. Ovarian weight was like that of control animals of the same age. Tertiary follicles and corpora lutea were demonstrable in the MP group. The histologic picture in the TP-treated animals differed radically from that of the controls or the other groups. Even after the 4-week pause we found only primary and secondary follicles and the weight increase was due to a luteinization of stroma.

The ovarian weight of CA- and DY-treated groups did not differ from the control groups at any time. But in MP-, DY- and CA-treated groups there was significant decrease in the number of corpora lutea and significant increase of tertiary follicles in MP- and DY-treated animals in the 8-week-old groups. After 4-week therapy pause the number of corpora lutea and tertiary follicles in DY- and MP-treated animals resembled that of the controls, while in CA-treated animals the number of corpora lutea and tertiary follicles was still reduced.

The sex steroids used affected the adrenal glands more intensely and for longer. The weight of the adrenal glands of the 8-week-old animals receiving MP and CA was less than that of controls. In contrast, those of the TP-treated animals were heavier. In the MP and CA groups the adrenals revealed a distinct reduction in the fascicular zone; in the TP group a stimulation, especially of the reticular zone. Both, the inhibitory effect and the stimulatory effect were still evident after the 4-week pause in treatment. DY treatment failed to affect the adrenal cortex either morphologically or gravimetrically.

Our studies showed that (a) perinatal treatment with TP prevents later cyclic ovulation, (b) synthetic gestagens with similar effects on the endometrium have different effects on the adrenal glands and ovaries, and (c) after 4-week pause in gestagen treatment the changes observed before in the ovary may have resolved, whereas in the adrenal gland the inhibitory or stimulatory action of the sex hormones may persist.

A number of investigations in the past (Zieger et al. 1974, 1975) have shown that sex steroids affect the adrenal cortex of the Syrian golden hamster. In the normal animal at puberty, the adrenal cortex begins to show sexual dimorphism, i.e., that of the male grows heavier than that of the female and develops a thicker cortex, particularly as a result of a broadening of the reticular zone. In the present study we investigate the effect and duration of effect of various synthetic gestagens and testosterone on the hamster ovary and adrenal cortex.

Groups of sixteen female hamsters were administered one of the following hormones: medroxyprogesterone acetate (MP), cyproterone acetate (CA), dydrogesterone (Dy), and testosterone propionate (TP). Administration was perinatal, beginning on the 4th day after birth.

The hamster brain, as we know, has not yet fully matured at this time. Perinatal administration of androgens prevents cyclic hormone secretion in the female, and leads to continual estrus and with it to lifelong sterility. The hormones were given in increasing doses, at first twice weekly and thereafter once a week, for a period of 8 weeks (Table 1).

Half the animals were killed at the end of the 8th week, and the remaining animals after a 4-week, therapy-free interval. Body weight, weight of ovaries and ad-

Table 1. Administration schedule

	4th day	7th day	11th day	3rd–5th week per week	6th–8th week per week
C Control, Sesame oil	0.025 ml	0.05 ml	0.05 ml	0.1 ml	0.1 ml
MP Clinovir	0.1 mg	0.2 mg	0.2 mg	0.4 mg	0.8 mg
DY Duphaston	1.25 mg	2.5 mg	2.5 mg	5 mg	10 mg
TP Testosterone propionate	0.25 mg	0.5 mg	0.5 mg	1 mg	2 mg
CA in Diane[a]	1.25 mg	2.5 mg	2.5 mg	5 mg	10 mg

a Cyproterone acetate (the gestagen component in Diane)

Table 2. Effects of sex steroids on Syrian golden hamsters, **a** following 8 weeks administration, **b** following 8 weeks administration and 4 weeks recovery period

a

	MP	Dy	CA
Ovary weight	↓	=	=
Number of tertiary follicles	↗	↗	↓
Number of corpora lutea	↓	↓	↓
Adrenal cortex weight	↓	=	↓
Number and size of cells of the fascicular zone	↓	=	↓

b

	MP	Dy	CA
Ovary weight	=	=	=
Number of tertiary follicles	↓	↓	↓
Number of corpora lutea	↗	=	↓
Adrenal cortex weight	↓	=	↓
Number and size of cells of the fascicular zone	↓	=	↓

renal cortex were determined. Cortex and glands were fixed in Bouin and examined under the light microscope (H & E, Masson-Goldner, PAS). The area of the ovaries on a histologic section was determined at 55x magnification using a MOP (semiautomatic electronic computer), and the tertiary follicles and corpora lutea counted in this area. Then this count was projected onto an area of 1 sqcm. The weights and other values were statistically evaluated using the Bartlett test, variance analysis, multiple *t*-test, parameter-free test of the Wilcoxon type, and Brand-Snedecor test (Snedecor and Cochran 1971).

After 8 weeks of hormone treatment particularly the TP-treated animals ($P < 0.001$), but also those given MP ($P < 0.05$) had ovarian weights that were significantly lower than those of the controls. The 4-week recovery period brought their ovarian weights back up to the control level (Fig. 1).

However, the histologic picture, in the TP-treated animals, differed radically from that of the controls. In both 8 and 12-week-old control animals we found follicles in all stages of maturation and a small number of corpora lutea. The TP-treated animals by contrast showed primary and secondary follicles almost exclusively. Though the ovary weights of these animals increased after 4 weeks of recovery, this increase was apparently not due to tertiary follicles and corpora lutea, as we might have expected, but to luteinization of stroma.

The number of corpora lutea decreased in all groups as a result of gestagen treatment. By contrast, the number of tertiary follicles increased in the MP and Dy ani-

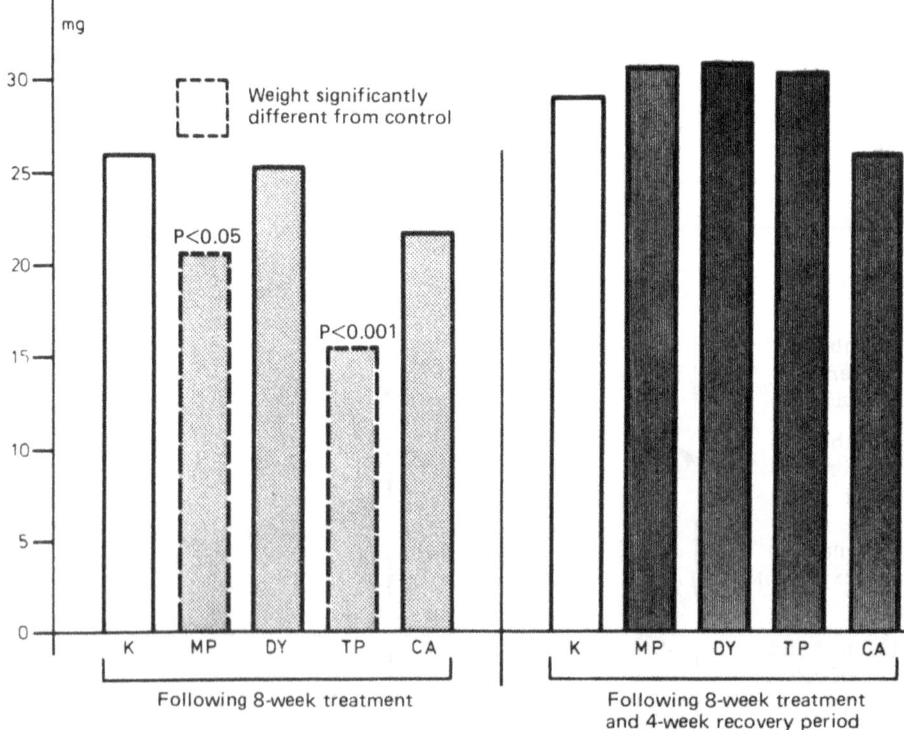

Fig. 1. Hamster ovarian weights

mals, and decreased in the CA group. After 4 weeks of recovery both MP and DY groups again had control levels of corpora lutea, but a lower number of tertiary follicles. The CA animals did not return to a normal level of either tertiary follicles or corpora lutea, i.e., their ovarian function still seemed to be suppressed (Fig. 2).

The adrenal cortex of the animals was also influenced in various ways by the 8-week hormonal treatment. Adrenal weight decreased significantly in MP ($P < 0.001$) and CA ($P < 0.001$) groups, and increased in TP animals ($P < 0.001$). Only the animals on DY had adrenal weights corresponding to those of the controls. The 4-week recovery period did not change this picture. Though all animals showed a significant growth effect compared to the younger animals on the same therapy, all groups still evinced a significant weight difference as against same age controls (Fig. 3):

MP 8 (MP treated animal, 8 weeks old): MP 12 (12 weeks old animal, treated with MP for 8 weeks) $P < 0.01$
Dy 8 (Dy treated animal, 8 weeks old): Dy 12 (12 weeks old animal, treated with Dy for 8 weeks) $P < 0.001$
TP 8 (TP treated animal, 8 weeks old): TP 12 (12 weeks old animal, treated with TP for 8 weeks) $P < 0.001$
CA 8 (CA treated animal, 8 weeks old): CA 12 (12 weeks old animal, treated with CA for 8 weeks) $P < 0.05$

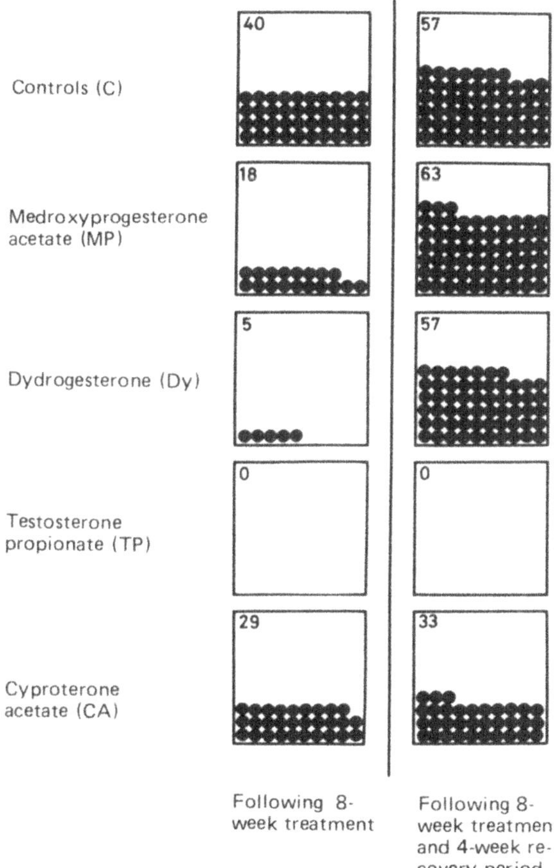

Fig. 2. Number of corpora lutea per cm^2

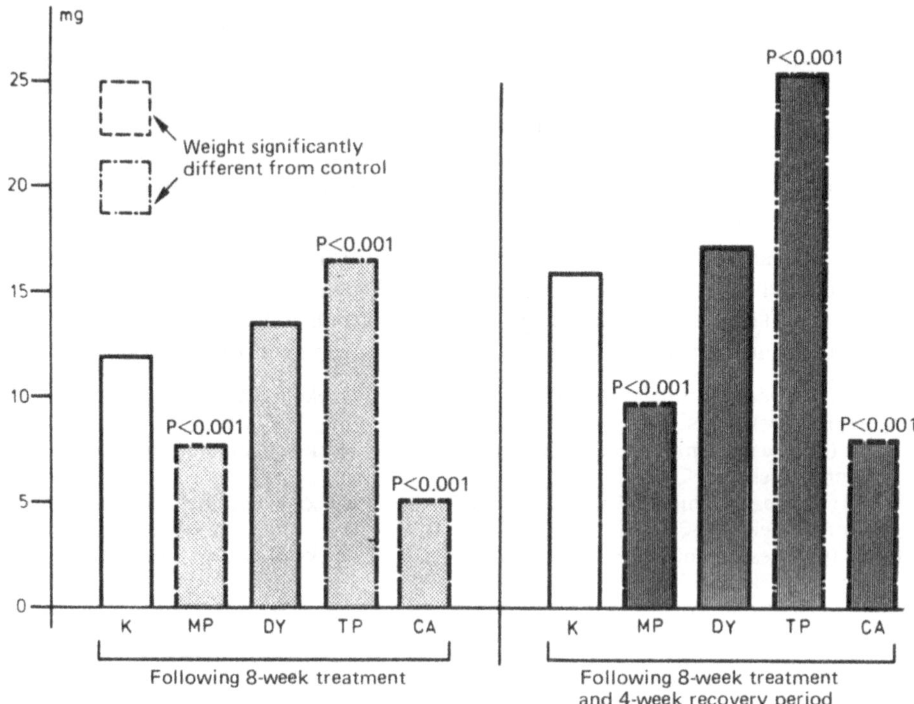

Fig. 3. Adrenal weight in female hamsters

In keeping with the known stimulatory effect of testosterone on hamster adrenal cortex, we saw an expansion of the cortex in treated animals as compared to controls; the cells of the outer fascicular zone were mitotic, a sign of a tendency to proliferate; and the reticular cells had developed beyond the stage seen in the controls.

DY seemed to have no effect on adrenal cortex. CA and MP, however, caused the fascicular zone to atrophy, an apparent glucocorticoid-like effect. Being a known anti-androgen, CA had a testosterone-like effect on reticular cells of the female hamster, similar to that described by Denef in 1968 on the enzyme activity of the rat liver.

The 4-week recovery period was not sufficient to reverse these characteristic changes in adrenal cortex (Table 5).

In sum we may say that synthetic gestagens, even if administered perinatally, lead to an only temporary suppression of ovarian function. They may affect ovary and adrenal cortex independently and in various ways. In general this effect is more prolonged on the adrenal cortex than on the ovary. Some synthetic gestagens have a glucocorticoid-like effect that can lead to atrophication of the adrenal cortex; this atrophication remains for long periods even after cessation of hormone therapy. This fact should be kept in mind when administering these substances to humans.

References

Denef C, Vandeputte M, de Moor P (1968) Paradoxical effects of the androgen antagonist cyproterone acetate as steroid metabolism in the rat. Endocrinology 83:945

Snedecor, Cochran (1971) Statistical methods, VIth edn, The Iowa State University Press, p. 128, 296–298

Zieger G, Lux B, Kubatsch B (1974) Sex dimorphism in the adrenal of hamsters. Acta endocr. (Kbh) 75:550

Zieger G, Breuner H, Kubatsch B (1975) Sex dimorphism of the adrenal cortex of the hamster following gonadectomy. Acta endocr. (Kbh) Suppl. 199:311

Clinical Effects of Gestagens

H. Wittlinger and C. Grumbrecht

Abstract

The physiologic effects of progesterone, which is produced in the organism, are presented. The endocrinologic, genital, extragenital, psychic, and neurovegetative effects of treatment with synthetic gestagens are also given, together with the different indications for gestagen therapy. Finally, the undesirable side effects of such therapy, and in particular those resulting from long-term usage, are discussed.

The only gestagen that occurs physiologically in the organism is progesterone, which is formed in the corpus luteum of the ovary. It induces endometrial transformation, insures nidation of the fertilized egg, brings the myometrium into pharmacodynamic quiescence, and finally enters the urine as the decomposition product pregnanediol. Since progesterone is ineffective when given orally, only partially soluble in oil, and is rapidly eliminated from the system, the search began early on for a substitute gestagen.

Gestagens affect the uterus, fallopian tube, ovary, pelvic connective tissue, mammary gland, integument, as well as the hormonal regulation system, the vegetative nervous system, the psyche, and the levels of water and minerals in the body. Finally, they cause a reduction in mitoserates, which is why they are used to treat certain types of tumor.

The indications for gestagen therapy in the field of gynecology and obstetrics are menstrual disorders, premenstrual syndrome, contraception, endometriosis, uterus myomatosus, functional sterility, mastopathy, imminent, abortion, endometrial carcinoma, ovarian carcinoma and mamma carcinoma. In the nongynecologic realm there are a number of indications for gestagen-containing drugs, among which are dermatologic disorders (loss of hair, sclerodermia, acne); hematologic disorders (anemia, illnesses requiring an anticoagulant treatment); oncologic disorders (cancer of the endometrium, breast, or prostata); and urologic disorders (e.g., prostata hypertrophy).

The side-effects of gestagens are less serious than those we see with estrogen treatment. Only when they are administered over long periods of time can they result in such disorders as loss of libido, depression, vaginal dryness, lessening of menstrual bleeding to the point of secondary amehorrhea, interim bleeding, and finally, the possibility of an increased risk of thromboembolism.

III. Combined Hormones

Morphological Changes in the Cervix and Vagina Induced by Combined Hormonal Medication

B. Czernobilsky

The first reports dealing with the effect of birth control pills on cervical histology appeared in the sixties. In these articles the similarities between the effects of birth control pills on the cervix and those of pregnancy were emphasized. Thus, the changes described were dilated endocervical glands and hypersection of mucus (Maqueo et al. 1966).

In contrast, what appeared to be genuine glandular hyperplasia under the influence of hormonal medication was first described by Taylor et al. (1967) and was called "atypical endocervical hyperplasia". Ten of their patients used combined medication, seven sequential and in two no details were available. Seven presented with cervical polyps and three had cervical erosions. The lesions were described as tightly packed, glandular, or tubular elements lined by flattened to cuboidal cells with little mucin. There were also areas in which the cells were present in a reticulated pattern or in solid cords. There were no mitotic figures. Nucleoli were not prominent. Stroma was compact or edematous with inflammation. These lesions were multicentric.

In Candy and Abell's (1968) series of adenomatous hyperplasia of the uterine cervix there were 12 patients culled from all their cervical biopsy specimens obtained from 1963 to 1967 that exhibited adenomatous or atypical gland cell hyperplasia. All of these patients had received one of the progestogens, usually in combination with a small amount of estrogen. Eleven of the patients received these drugs for contraceptive purposes and one, aged 66, for carcinoma of the endometrium. The lesions were similar to those described by Taylor et al. (1967). The hyperplastic cells appeared to arise from both surface columnar epithelium and subcolumnar reserve cells. The predominant pattern was that of closely packed, small, glandular spaces formed by cuboidal cells poor in mucus with little intervening stroma. Nuclei were uniform with very few mitoses.

Kyriakos et al. (1968) described 30 patients, of whom 22 had been receiving oral contraceptives on a cyclic basis for 2 weeks to 6 years with a mean of 20.7 months, three patients were pregnant and two received hormone injections. Grossly the cervix showed erosions and polyps in four patients. These authors described two histologic lesions: (1) Squamous metaplasia of the endocervix with persistence of mucin producing glands within the new squamous epithelium also called squamocolumnar prosoplasia; and (2) microglandular hyperplasia consisting of crowded glands which varied in size and showed some subnuclear mucus containing vacuolization. The epithelium lining these glands was cuboidal or flattened, and there was no nuclear atypism and no mitotic activity. The stroma between these glands often showed acute and chronic inflammation. These authors concluded that while squamous metaplasia with entrapped mucin producing glands can be seen with either pregnancy or oral contraceptives, microglandular hyperplasia is more characteristic of contraceptive pills. In 13 patients cervical smears were available and these showed nonspecific inflammatory changes. Kyriakos et al. (1968) concluded that this hyperplastic and cervical lesion is definitely related to combined hormonal medication. They also suggested that the lesion proceeds from squamous prosoplasia to mixed squamous-glandular proliferation to pure microglandular hyperplasia.

Gall et al. (1969) considered many different aspects of cervical reaction to the birth control pills. They reviewed 103 randomly selected patients attending a family planning clinic who had been taking oral contraceptives for periods ranging from 1 to 60 months and found lesions not staining with Schiller solution in 84% of the patients. The squamous epithelium showed increased vascularity, infiltration by neutrophiles, presence of mitoses, an increase in number of cells above the basal layer, some dysplasia, and intraepithelial vesicles. The stroma was inflamed and edematous. Only in 3 cases out of the 103 examined did they find the typical adenomatous endocervical hyperplasia. Gall et al. (1969) concluded that reactivity of the cervix to contraceptive agents increases with the length of their use and that combination agents caused more changes than sequential ones.

Govan et al. (1969) suggested that these endocervical lesions arise only in preexisting polyps, which also presented reserve cell hyperplasia.

In a study by Nichols and Fidler (1971) microglandular hyperplasia was looked for in 128 consecutive cone biopsies taken because of suspicious or positive cytologic findings. Microglandular hyperplasia was found in 24% of the total series and in 44% of those with histories of oral contraceptive use. Of the patients who had not taken pills, only 11% had similar lesions and half of these were pregnant, postpartum, or with unknown contraceptive histories. The high incidence of the hyperplastic features in their series may have been due to the fact that serial sections were examined and that tiny foci of glandular hyperplasia were also included as positive finding. Some of the patients in this series showed persistence of the lesions although having stopped medication for some time. This suggests that glandular proliferation, once initiated by progesterone stimulation will last for some time even after withdrawal of the drug. There was no significant correlation with age, cytologic atypia, or cone biopsy diagnosis.

Wilkinson and Dufour reviewed the subject in 1976. Out of their 22 cases 6 were not associated with hormonal changes. The length of hormonal therapy had no impact on the histologic lesions. One of their patients also had a squamous cell carcinoma in situ of the cervix. Histologically the lesions were similar to those described by the above-mentioned authors. These authors considered most of the cells filling the spaces between the hyperplastic glands to represent reserve cell type epithelium. They concluded that the overall incidence of cervical microglandular hyperplasia among patients taking the pill ist between 2%–3%.

Because of the microscopic appearance of microglandular hyperplasia, an erroneous diagnosis of malignancy can sometimes be reached, but if one remembers that in the benign hyperplastic lesions there is very little nuclear atypism, only rare mitoses, and no invasion, such an error can be avoided. Microglandular hyperplasia also has to be differentiated from clear cell carcinoma of the cervix. While the glands of microglandular hyperplasia are uniform and small, those of clear cell carcinoma vary in size, are mixed with sheets of tumor cells showing marked nuclear pleomorphism, prominent nucleoli, clear cells, and hobnail cells. In addition, the cytoplasm of the clear cell carcinoma contains glycogen and no mucin, while in microglandular hyperplasia glycogen is absent and some mucin is usually present within the cells. The glandular lumens of both conditions may contain mucin positive material.

Endocervical malignancy arising from microglandular hyperplasia has, to the best of my knowledge, never been demonstrated. On the other hand there are now in the literature reports of four patients, including one from our laboratory, who developed cervical adenocarcinoma or mixed adenosquamous carcinoma after having been on birth control pills for some time (Czernobilsky et al. 1974). In these cases as well, no transition from benign to malignant lesions was shown and thus the relationship between the pill and the cervical malignancies in these patients may very well have been entirely fortuitous. Nevertheless vigilance in this area should be practiced. Patients with microglandular hyperplasia should be followed after cessation of hormonal therapy and if necessary rebiopsied at a later date.

Vaginal microglandular hyperplasia is extremely rare and occurs only in those patients who also have vaginal adenosis. Robboy and Welch (1977) reported eight examples of microglandular hyperplasia that arose in areas of vaginal adenosis. Five of the patients with both microglandular hyperplasia and vaginal adenosis of endocervical type took oral contraceptives and had also been exposed prenatally to DES. The distinction between these two lesions is important since clear cell carcinoma can arise on a background of vaginal adenosis, while as I mentioned before has not been observed to originate in microglandular hyperplasia.

In conclusion, microglandular hyperplasia of the cervix is characterized by crowded often tightly packed, glands usually presenting so-called reticulated or solid pattern. The epithelial lining cells are quite uniform with very little nuclear pleomorphism and no to rare mitoses (Figs. 1–3). This lesion occurs predominantly in women on combined birth control medication and appears to be due to progestogens. It differs from the endocervical glands seen in pregnancy where the glands are markedly enlarged with prominent hypersecreting mucus cells. It has been estimated that in the cervix this lesion occurs in about 2% to 3% of pill users, but accurate incidence studies

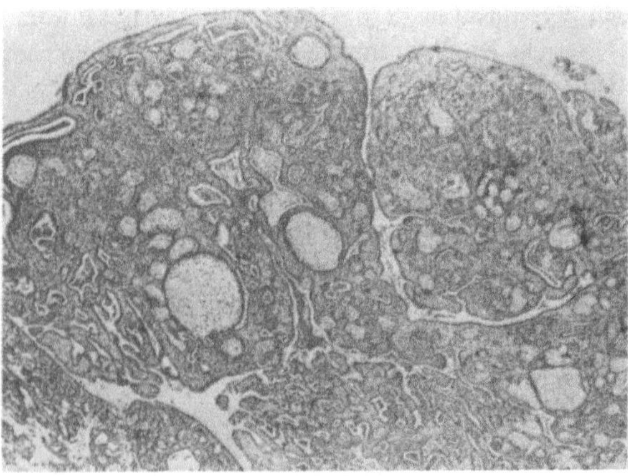

Fig. 1. Polypoid lesion of endocervix showing predominantly small irregular glands with marked stromal and intraluminal inflammatory infiltration, in patient on combined birth control medication. H and E stain. x 16

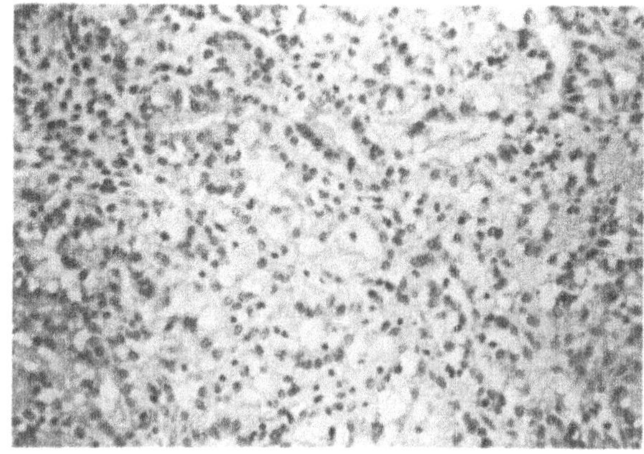

Fig. 2. Microglandular hyperplasia showing solid, so-called reticulated growth of cervical glands with some mucus vacuoles in endocervix of patient on combined birth control medication. H and E stain. x 100

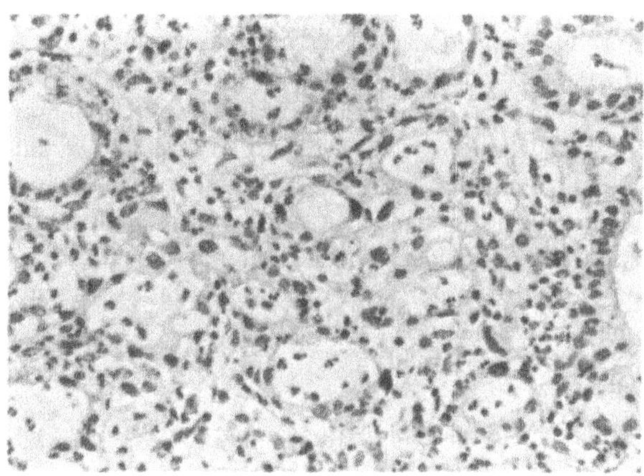

Fig. 3. Higher magnification of endocervical microglandular hyperplasia showing nuclear atypism but no mitoses. H and E stain. x160

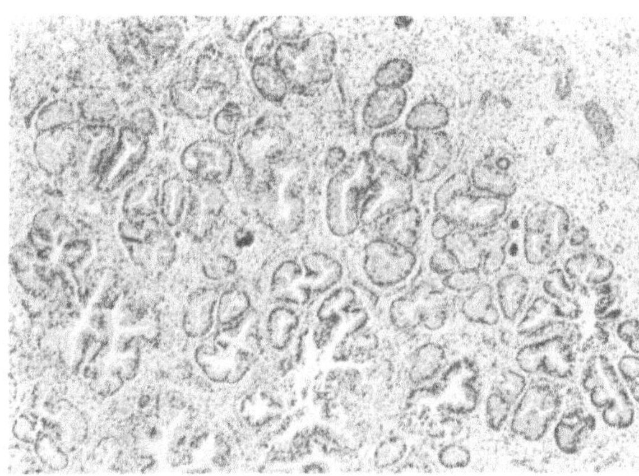

Fig. 4. Less typical form of endocervical glandular hyperplasia characterized by relatively large but not particularly crowded mucus rich glands in patient who did not receive any hormonal medication. H and E stain. x 24

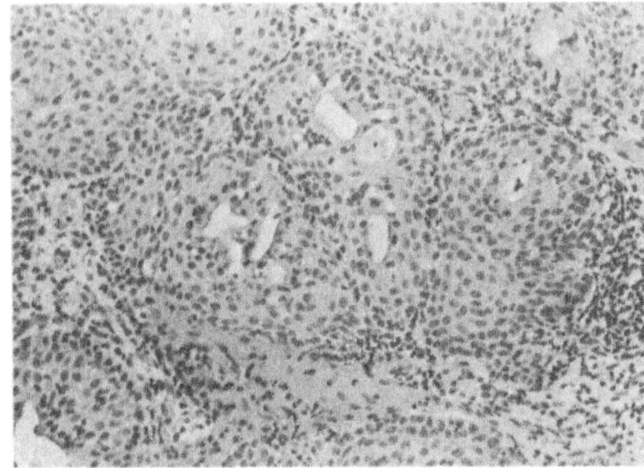

Fig. 5. Marked endocervical squamous metaplasia with entrapped mucus glands (squamocolumnar prosoplasia). This finding has been reported in patients on birth control medication, during pregnancy and in neither condition. H and E stain. x 100

are still lacking. Occasionally the lesion can also be seen in women who receive no hormonal therapy and are not pregnant. In these patients the glandular hyperplasia is usually composed of less crowded, irregular endocervical glands of tubular type (Fig. 4). Squamous metaplasia with or without entrapped glands is of no particular help in distinguishing patients on the pill from pregnant women or from those not exposed to either condition (Fig. 5). In the vagina microglandular hyperplasia is extremely rare and only occurs in the presence of vaginal adenosis from where it arises. Microglandular hyperplasia must be differentiated from adenocarcinoma and clear cell carcinoma. To the best of our knowledge the lesion is not premalignant and so far no malignant transformation has been demonstrated. In most instances the lesion disappears after cessation of the hormonal therapy but follow-up of these patients and in case macroscopic features persist, rebiopsy may occasionally be indicated.

References

Candy I, Abell MR (1968) Progestogen-induced adenomatous hyperplasia of the uterine cervix. JAMA 203:323–326

Czernobilsky B, Kessler I, Lancet M (1974) Cervical adenocarcinoma in a woman on long-term contraceptives. Obstet Gynecol 43:517–521

Gall SA, Bourgeois CH, Maguire R (1969) The morphologic effects of oral contraceptive agents on the cervix. JAMA 20:2243–2247

Govan ADT, Black WP, Sharp JL (1969) Aberrant glandular polypi of the uterine cervix associated with contraceptive pills: Pathology and pathogenesis. J Clin Pathol 22:84–89

Kyriakos M, Kempson RL, Konikov NF (1968) A clinical and pathologic study of endocervical lesions associated with oral contraceptives. Cancer 22:99–110

Maqueo M, Azuela JC, Calderon JJ, Goldzieher JW (1966) Morphology of the cervix in women treated with synthetic progestins. Am J Obstet Gynecol 96:994–998

Nichols TM, Fidler HK (1971) Microglandular hyperplasia in cervical cone biopsies taken for suspicious and positive cytology. Am J Clin Pathol 56:424–429

Robboy SJ, Welch WR (1977) Microglandular hyperplasia in vaginal adenosis associated with oral contraceptives and prenatal diethylstilbestrol exposure. Obstet Gynecol 49:430–434

Taylor HB, Irey NS, Norris HY (1967) Atypical endocervical hyperplasia in women taking oral contraceptives. JAMA 202:185–187

Wilkinson E, Dufour DR (1976) Pathogenesis of microglandular hyperplasia of the cervix uteri. Obstet Gynecol 47:189–195

Current Concepts in Steroid Hormone-Cell Interactions in the Uterus and Chick Oviduct

E.E. Baulieu and R. Mortel

Abstract

This paper gives a summary of the findings which have led to the present understanding of steroid hormone receptors. It discusses a new physiologic approach to receptor pharmacology, based on receptor regulation, differential hormone binding properties and cellular plurality of receptors.

Specific Steroid Hormone Binding Proteins in Target Cells Are "Receptors"

The *estradiol receptor* (ER) has been a "pioneer" not only in the field of steroid receptors but also with respect to the receptors of other hormones and neurotransmitters. The concept of a receptor was originally defined by pharmacologists who, however, only studied them at the phenomenological level. It was not until high specific activity [^{3}H]estradiol was available that endocrinologists and biochemists were able to study the physico–chemical interactions.

The natural estrogen, estradiol, is *retained* in its target organs (Jensen and Jacobson 1962). This basic observation was explained by the *intrinsic* properties of an intracellular protein component of target cells when a *strict specificity*, of obvious biologic significance, was demonstrated by in vitro binding experiments with soluble extracts (Baulieu et al. 1967; Toft et al. 1967). Two techniques were instrumental in this: (1) *gradient ultracentrifugation* of low salt cytosol extracts permitted the isolation of "8S" [^{3}H]hormone complexes (Toft and Gorski 1966); (2) the rational use of adsorbents (charcoal in the first place), according to the *differential dissociation* principle (Milgrom and Baulieu 1969), allowed quantitative binding studies (Baulieu et al. 1967). The steroid receptors show *highest affinity* for the most biologically active hormones with K_{Deq} values on the order of 0.1 nM, which relate to the circulating concentrations of these hormones. The very slow dissociation rates of the hormone-receptor complexes (Best-Belpomme et al. 1970; Truong and Baulieu 1971) are compatible with the long retention observed in vivo (Jensen and Jacobson 1962).

On the basis of these properties, despite only circumstantial evidence that these *intracellular* binding proteins actually operate in mediating hormone action (Baulieu 1973), they were called receptors. They are certainly *distinct* from the other steroid binding proteins found in plasma, testicular fluid, liver, prostate, etc., whose binding properties do not correlate with the biologic activities of their ligands (Baulieu 1973). For instance, human sex steroid binding plasma protein (SBP) (Mercier-Bodard et al.

1970), which binds estradiol tightly, binds testosterone with greater affinity, in contrast to ER which has very weak affinity for androgens. Conversely, diethylstilbestrol (DES), an active synthetic nonsteroidal estrogen, binds to ER but hardly at all to SBP or rat α-fetoprotein (α-FP), also called estradiol binding protein (Raynaud et al. 1971). Recent findings with *antibodies* to ER (Fox et al. 1976), α-FP (Radanyi et al. 1977), and SBP (Renoir et al. 1977) confirm that the intracellular ER is distinct from the plasma proteins binding estradiol, and show that the ERs of different organs in different mammals have common antigenic determinants (Greene et al. 1977).

For each natural hormone, binding experiments have essentially indicated *a* single category of binding sites of high affinity (K_{Deq} = 0.1 nM) – no cooperative binding has been indicated. No work based on purification, antigenicity, or binding activity has substantiated the notion of receptor heterogeneity. The frequently referred to "heterogeneity" of receptor, which corresponds mainly to differences in size or extractibility, may therefore be artifactual (due to aggregation, proteolysis). The significance of the "insoluble" nuclear receptor (Lebeau et al. 1973) and the reason why a fraction of progesterone binding sites do not translocate to the nucleus (Mester and Baulieu 1977) are still obscure. Possibly R can display distinct interactions with several structures of the same cell. Moreover, R may interact selectively with different acceptors and/or effectors in different cells and this could be the basis for the same hormone giving rise to different effects via the same receptor.

The structures of the steroid receptors are not known, despite much progress in affinity chromatography techniques (Sica et al. 1973; Kuhn et al. 1975; Truong et al. 1973). We only know that they are *proteins* (mol. wt. approx. 70,000 in the case of the calf uterus ER binding units), antigenic whether extracted from the cytosol (Fox et al. 1976) or nucleus (Greene et al. 1977), and that partial proteolysis removes a portion of the protein without changing the hormone binding characteristics (Truong et al. 1973; Erdos 1968; Wrange and Gustafsson 1978).

Steroid Hormone Receptor Complexes Act in the Cell Nucleus

Steroid hormone receptor complexes concentrate in the nucleus of target cells, where presumably they trigger the main hormonal effect(s) by modifying gene transcription.

It has been kown for a long time that many responses elicited by steroid hormones implicate *changes in protein and RNA synthesis* (Mueller et al. 1958). Recent evidence demonstrates an accumulation of specific mRNA(S) when specific protein synthesis is increased (O'Malley and Means 1974; Schimke et al. 1977). That receptors are implicated in these events has been suggested by two series of observations.

Firstly, in most systems the receptor is cytoplasmic in the absence of hormone and *translocates* to the *nucleus* with the bound steroid after exposure to the hormone (Jensen et al. 1968; Gorski et al. 1968). This change in subcellular distribution apparently reflects *activation* of the receptor – a steroid-dependent phenomenon increasing its affinity for polyanions and DNA in particular (Higgins et al. 1973; Milgrom et

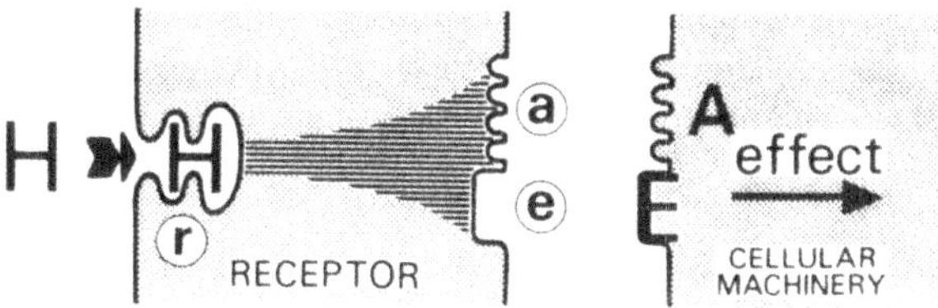

Fig. 1. Phenomenological portrait of steroid hormone (H) receptor (R). *r* is the *binding site of H*, and can stand for "receptor", "receiving", "regulatory". r is hormone *specific*, binds *reversibly* and is the *last* interaction of hormone with the cellular machinery before action is triggered (this would not be the case if r was the binding site of a transport protein). The receptor includes a *coupling mechanism* (transduction) between r and site (s) and/or e. The effect takes place at the level of the effector structure *E* of the cell machinery. The receptor has a *site e* (for "effect", "effector", "executive") interacting with E, either catlyzing a chemical reaction (R is then an allosteric enzyme) or simply binding to E (in that case, sites e and a may be confounded). In both cases, some specificity should be present to explain that different effects are promoted by different steroid-receptor complexes in the same cell. *A* is a (another) part of the cellular machinery called *Acceptor* to which the *a site* of the receptor can bind. The effect of an *antihormone* binding to the r site may result in changes in the conformation of the r site, consequently modifying transduction to the a and/or e sites, and thus abolishing hormone effect

al. 1973a).[1] Autoradiography studies have shown the preferential localization of steroids in target cell nuclei (Edelman et al. 1964) and, more recently, a selective association of receptor with solubilized chromatin elements has been described (Massol et al. 1978). However, it will be seen later that nuclear transfer of the receptor after ligand binding does not necessarily lead to further action. Thus there is some *distinction between acceptor and effector* sites of R, as depicted in Fig. 1. In addition translocation of the receptor is probably not a prerequisite for hormone action in all cases (i.e., when receptor is already located in the nucleus in absence of hormones (Mester and Baulieu 1972; Zava and MacGuire 1977) or when previously translocated by antihormone (see part 4).

Secondly, *increased incorporation* of radioactive-*RNA* precursors has been observed when nuclei of target cells have been exposed to cytoplasmic extracts containing hormone-receptor (Raynaud-Jammet and Baulieu 1969). This observation gave some evidence for the *direct* participation of the receptor (hormone alone has no effect), although knowledge of eukaryotic transcription at that time was, and still is, very primitive. So far *no unequivocal* evidence has been obtained which demonstrates that the addition of steroid receptor to a chromatin preparation can result in a change (for instance an increase in specific mRNA) mimicking the response observed in the intact cell (Schwartz et al. 1976). Indeed all steroid responses are poorly understood mechanistically. For instance, inhibition of protein synthesis has no effect on the increase in viral RNA by glucocorticosteroid in GR cells (Ringold et al. 1977), in con-

1 The further fate of hormone receptor complexes in the living cells is not known. "Deactivation" of R or alteration by proteolysis are amongst the possibilities that have been proposed

trast to the induced early augmentation in conalbumin and ovalbumin mRNAs (MacKnight 1978). Moreover, the increase in conalbumin (and $mRNA_{con}$) occurs very early after estradiol administration, and much later after progesterone, whereas the increase se in ovalbumin (and $mRNA_{ov}$) due to estradiol occurs later than that due to progesterone, but before the effect of the latter on conalbumin (Schimke et al. 1977; Palmiter et al. 1976). These kinetics, observed in the same chick oviduct tubular cells, cannot be explained by current knowledge, and the concept of a "simple" interaction of the receptor(s) with some segment(s) of the DNA encoding a regulatory signal for transcription, therefore seems less tenable.

Steroid Hormone Receptor Concentrations Vary in Target Cells and this Plays an Essential Role in Hormonal Receptivity

Steroid hormone receptor concentrations can vary under diverse hormonal, developmental, genetic, pathologic, and pharmacologic circumstances. These variations in spite of the limited information on structure and molecular mechanism of activity of receptors, are important in the understanding of all hormonally controlled processes. Endocrinology should therefore include the study of not only hormone but receptor variations.

Hormone-induced variations were the first to be described in detail, and the study of the uterine progesterone receptor during the estrus cycle led to discovery of *the double regulatory control* of steroid receptor concentrations (Milgrom et al. 1973b). A *positive* control is readily observed when, under physiologic and pharmacologic conditions, progesterone receptor synthesis is increased by estradiol – a result probably explaining the *priming* effect of estrogen on progesterone action. Such a positive control is more difficult to analyze when it is the hormone itself which increases the concentration of its receptor, as for estradiol and androgens. Inversely, progesterone exerts a *negative* control by accelerating the decay of its available receptor binding sites (Milgrom et al. 1973b) – an "inactivation" process which is not fully understood at the molecular level. "Down regulation", similar to the progesterone effect, then has been described for many polypeptide hormone receptors (Kahn 1976) and has attracted much attention since it has been correlated functionally with the desensitization of target cells and physically with an "internalization" process.

Positive and negative controls of receptor concentration are not the same according to the cell type, state of development, hormonal, metabolic, and functional circumstances. Such a *complexity* is obvious when one studies the interplay between progesterone and estradiol actions, since besides the above-mentioned data, it has been observed that progesterone may increase estradiol receptor concentrations directly or decrease the estradiol-dependent increase in estradiol receptor (Mester et al. 1974; Hsueh et al. 1975; Tseng and Gurpide 1975), etc. A remarkable sequence of variations is found in the *human endometrium* during the normal menstrual cycle (Bayard et al. 1978) (Fig. 2). Subcellular distribution and changes in concentration vary greatly but apparently follow the same rules as established for laboratory animal model systems. Incidently, the concentration of ER and its control are not identical in two neighboring tissues of the uterus: endometrium and myometrium (Mester et al. 1974; Alberga and Baulieu 1965; Jensen and de Sombre 1972).

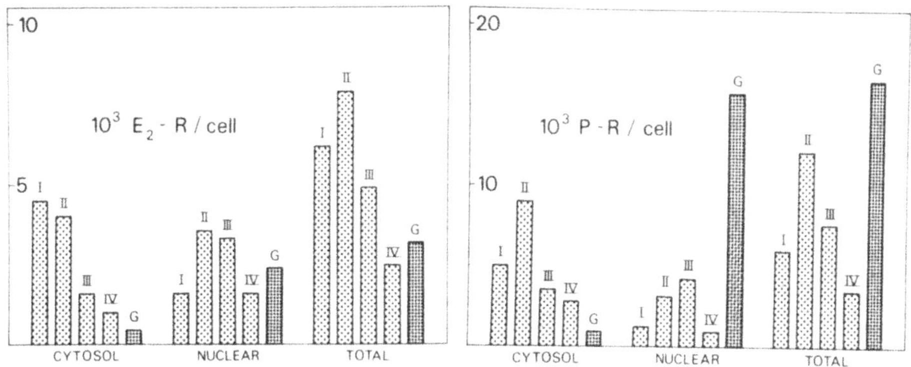

Fig. 2. ER and PR in the cytoplasm and the nuclei of human endometrium during the normal menstrual cycle divided into four periods with respect to the Lh peak (day 0); *I*: before -5, *II*: between -5 and 0, *III*: between 0 and 5, and *IV*: after +5. *G* are pregnancies (6-8 weeks). *Total* = cytosol + nuclear. (After Bayard et al.; Robel et al.; and Verma, unpublished information)

Changes in receptor concentration accompany not only functional variations but also *ontogenic* development. Timely programmed changes in androgen receptivity of the prostatic (Jost 1961) or mammary (Kratochwil and Schwartz 1976) buds are not under hormonal control, whereas other developmental changes are hormone-dependent, e.g., the peak of estrogen sensitivity (Le Goascogne and Baulieu 1977) and ER concentration (Gorski et al. 1971) in the rat uterus coincide with plasma estrogen increase at 10 days. However, the developmental curves of ER concentrations in other organs, such as the hypothalamus and pituitary, do not follow the same pattern (Raynaud and Moguilewsky 1976).

The variation in steroid receptor concentrations is also important in *clinical* situations. In human breast (Jensen et al. 1971; Mac Guire et al. 1977) and endometrial (Mac Guire et al. 1977; Robel et al. 1978) *cancers*, one observes a wide range of concentrations for the estradiol and the progesterone receptors, unrelated to the known histologic and etiologic particularities of these tumors. Correlation between responses to endocrine treatment and a selected threshold concentration of ER have been proposed (Jensen et al. 1971), although possible cellular heterogeneity makes interpretation and prognosis difficult. Nonetheless it was clearly shown that in the absence of detectable receptor or at concentrations below the selected threshold, the tumors show no hormonal response.

Remarkably, whatever the cause of the changes in receptor concentration, whether functional, ontogenic, genetic, pathologic, or pharmacologic (as seen in the next section), the binding specificity and affinity of the receptors has always been the same. This implies that one is dealing with an apparently purely *quantitative* regulatory system. However, this relatively simple approach does not explain all observed effects and the relationship between the concentration of hormone-receptor complexes and the extent of the response depends on the one considered. In the chick oviduct, for instance, half maximal induction of ovalbumin necessitates approximately twice the hormone (whether estradiol or progesterone) needed to obtain half maximal conalbumin increase (Palmiter et al. 1978). According to the dose, estradiol and progesterone can be agonistic or antagonistic for the same parameter (ovalbumin synthesis) and under certain circumstances, progesterone may be simultaneously synergic to estradiol for ovalbumin and conalbumin synthesis and antiestrogenic with respect to the increase in DNA polymerase (Sutherland et al. 1977a). Several hypothetical mechanisms could be responsible for these situations. For instance different numbers of the same receptor may be engaged in the hormone regulation of different gene products or there may be as yet unknown intermediary step(s) between the receptor and the measured responses.

In conclusion, *presence* of the receptor is *necessary* for steroid hormone responses and its *concentration* is *directly implicated* in the extent of the effect obtained. The concentration of *nuclear* hormone receptor complexes and its persistancy is related to the extent of the effect obtained (Anderson et al. 1974; Katzenellenbogen and Gorski 1972; Sutherland et al. 1977b). Refinements of these notions should take into account the stereochemical and kinetic parameters of the interactions between hormones and receptors brought to light by binding studies (discussed further on), and the postreceptor elements of the cellular machinery involved in the response (Croce et al. 1974).

Physiologic Pharmacology of Natural and Synthetic Hormones and Antihormones, Based on the Binding Properties, Plurality, and Changes in Concentration of Receptors

Binding Properties and Plurality of Receptor

A given hormone may interact with *several* receptors. It was originally demonstrated that estradiol interacts with AR (Jung-Testas et al. 1976), with a K_{Deq} approx. 3 n*M* while estradiol binds to ER with K_{Deq} 0.1 n*M*. The ER and AR were found together in cloned cells and when estradiol is added, it apparently exerts a strong negative signal at the AR level overriding the androgen effect (Feyel-Cabannes et al. 1978; Jung-Testas and Baulieu 1979). Conversely the binding of androgens to ER, also difficult to demonstrate physically because of very weak affinity, is well documented, since testosterone and DHT at pharmacologic doses provoke estrogenic activity (Rochefort and Garcia 1976).

Progesterone provides another example since, besides its own receptor, it binds to AR, GR, and MR (Mac Guire et al. 1977). Progesterone is a weak androgen in prostatic cells (and moreover a progesterone derivative, cyproterone acetate, is an antiandrogen), and is also an antiglucocorticosteroid in hepatoma and lymphoid cells, and an antimineralocorticosteroid in kidney. This suggests that a natural hormone such as progesterone may be active in cells through different receptors at the same time. This is the case in MCF seven human mammary cancer cells where progesterone can find three direct chanels for receptor interaction with the cell machinery (PR, AR, and GR) plus, probably, a fourth indirect one via the regulation of ER concentration (Horwitz et al. 1975). In fact it is conceivable that a steroid can be active in a cell even in the absence of its own receptor. Such theoretical considerations indicate the difficulty of predicting the therapeutic value of a given hormone drug, even in the case of "natural" steroids.

If several different receptors can bind the same hormone, it follows that *a given receptor can bind several hormones.* The heterologous hormone will exert either an agonistic effect with respect to the receptor's own steroid, or a different effect, possibly antagonist no prediction being possible from binding properties alone.

Changes in Receptor Concentration

As demonstrated previously for the PR increase by estradiol, these may also be important pharmacologically. A new example has been obtained in endometrial cancer which is sensitive to progestagen therapy, and where we find PR in all cases (Robel et al. 1978). Since the treatment becomes inefficient with time and the progesteronelike compounds "inactivate" the PR (Milgrom et al. 1973b), it is of interest to note that tamoxifen (a triphenylethylene derivative with antiestrogenic properties) can increase significantly PR concentration in the cancerous tissue in vivo, even though the compound is not a strong estrogen by other criteria (Robel et al. 1978).

Antihormones

The principles applied to hormones also apply to antihormones, which antagonize hormone action at the target cell level. Again we can envisage two types of mechanisms.

Decreases in receptor *concentration* should diminish hormone activity. Based on this and the fact that progesterone "inactivates" its own receptor, we have proposed a method for "midcycle" contraception in which a ligand having little or no progestational activity prematurely inactivates PR, which is highly increased at this period of the cycle (Baulieu 1975). In the future, drugs decreasing selectively the concentration of hormone receptors may be of great use.

Nevertheless at the present time, one should simply envisage the antihormones as showing competitive *binding* to the receptor site of the hormone which they antagonize. The simplest and best antagonist would be a compound binding with enough affinity to exclude efficiently the hormone, a metabolism slow enough that it is available for a long period, and, last but not least, a structure such that its interaction with the receptor binding site does not lead to any hormonal agonistic effect. In other words, a "pure" antihormone of great activity. Such requirements have been met by *tamoxifen*, an antiestrogen, for which no estrogenic effect but a complete estrogen antagonism have been demonstrated in the chick oviduct system (Sutherland et al. 1977c), thus providing a model for understanding antihormone action. Preliminary indications show that its rate of association with the receptor is relatively slow and its rate of dissociation is rapid and the overall affinity rather high (Geynet, unpublished). Its slow metabolism is responsible for its lengthy availability to the cell. After binding to the receptor, "activation" takes place in the sense that there is transfer of tamoxifen-receptor complexes to the nucleus, where they stay for a long time while no estrogenic activity occurs. Ten milligrams of tamoxifen are required to abolish the effects of 1 mg of estradiol (actually a very high dose difficult to antagonize). Tamoxifen, when given after estrogen has been administered, stops hormone action (Mester et al. 1977), and conversely, estradiol given to an animal receiving tamoxifen can trigger estrogenic effects (Mulvihill and Palmiter, personal communication). Several important conclusions can be derived from these observations. The first is that hormone or antihormone action is not critically dependent on cytoplasmic receptor concentration, since tamoxifen and estradiol are active even under circumstances where the cytoplasmic receptor is low and most of the receptor is located in the nucleus. The second is that the cytoplasmic receptor is not necessarily required to allow a ligand to enter the nucleus. Thus the data suggest that the receptor is decisive for "interpreting" the molecular structure of the ligand, since the estradiol-receptor complexes are active and the tamoxifen-receptor complexes are inactive in terms of estrogenic activity, and that the importance of the receptor is not in terms of an intracellular transport proteins.

Although, tamoxifen has provided us with a relatively clearcut example, the problem is not often as simple with other so-called antihormones. Even tamoxifen is very difficult to understand in systems other than chick oviduct since it (and triphenyl ethylene congeners) displays mixed agonist and antagonist properties (Clark et al. 1974). In the central nervous system (in the rat), the antihormone (nafoxidin) is antiestrogenic for one behavioral test (lordosis) and estrogenic for another (eating) (Wade and Blaustein 1978). We have seen that in endometrial cancer, tamoxifen represents a "partial" estrogen giving potentially increased receptivity to progestational agents (Robel et al. 1978).

Although details of the integrated interactions between ligand, receptor and effector structure(s) will probably remain hidden for a long time, it is already apparent that the studies of ligand-receptor binding kinetics and of the availability of ligand to the target cells can help in the selection of better drugs. It is known that "weak" agonists of low affinity, due to fast dissociation rates, display antagonists effects when administered together with the agonist. However repeated administration will lead to full agonist activity, as demonstrated in the case of estriol vs estradiol (Huggins and Jensen 1955). Thus, with well chosen doses and an appropriate rhythm of administration, one can obtain hormone *or* antihormone activity as a consequence of the dynamics of binding, the metabolic characteristics, and/or conformational changes of the receptor.

Conclusions

The existence of specific intracellular hormone receptors in somatic target cells and their mediation of hormone action via gene transcription is now well established. Further understanding of steroid hormone mechnisms of action, at the molecular level, will benefit from purification of the respective receptors and immunologic studies with specific receptor antibodies. However, this approach is likely to be hampered by the current limited understanding of chromatin interactions and gene functioning, and progress in this field is therefore difficult to estimate.

The observations that several hormones may bind to one receptor, that several receptors can interact with the same hormone, that several receptors may be present in the same cell, and that receptor concentrations vary under different conditions, have led to the development of a new physiologic approach to pharmacology. This approach is readily applicable to synthetic agonists and antagonists (antihormones) and should permit progress in both applied (ie. diagnosis and therapy) and theoretical terms.

Acknowledgements

Grants from the Cnrs, the Dgrst, the Ford Foundation, WHO and Roussel-Uclaf are gratefully acknowledged. We also wish to thank Heather Mullis, Eileen Mulvihill, Francoise Boussac and Anne Atger for their help with this manuscript. The names of our colleagues appear in the appropriate references.

References

Alberga A, Baulieu EE (1965) Concentration élective de l'oestradiol dans l'endomètre chez la ratte. C R Acad, Sci [D] (Paris) 261:5226–5228

Anderson JN, Peck EJJr, Clark JH (1974) Nuclear receptor estradiol complexe: a requirement for uterotrophic responses. Endocrinology 95:174–178

Baulieu EE (1973) A 1972 survey of the mode of action of steroid hormones. Excerpta Med Int Congr Ser 273:30–62

Baulieu EE (1975) Antiprogesterone effect and midcycle (periovulatory) contraception. Eur J Obst et Gynecol Reprod Biol 4:161–166

Baulieu EE, Alberga A, Jung I (1967) Récepteurs hormonaux. Liaison spécifique de l'oestradiol à des protéines utérines. C R Acad Sci [D] (Paris) 265:354–357

Bayard F, Damilano S, Robel P, Baulieu EE (1978) Cytoplasmic and nuclear estradiol and progesterone receptors in human endometrium. J Clin Endocrinol Metab 46:635–648

Best-Belpomme M, Fries J, Erdos T (1970) Interactions entre l'oestradiol et des sites récepteurs utérins. Données cinétiques de l'équilibre. Eur J Biochem 17:425–432

Clark JH, Peck EJ Jr, Anderson JN (1974) Oestrogen receptors and antagonism of steroid hormone action. Nature 251:446–448

Croce CM, Koprowski H, Litwack G (1974) Regulation of the corticosteroid inducibility of tyrosine aminotransferase in interspecific hybrid cells. Nature 249:839–841

Edelman IS, Bogoroch R, Porter GA (1964) On the mechanism of action of aldosterone on sodium transport: the role of protein synthesis. Proc Nat Acad Sci USA 50:1169–1176

Erdos T (1968) Properties of a uterine oestradiol receptor. Biochem Biophys Res Commun 32:338–343

Feyel-Cabannes T, Secchi J, Robel P, Baulieu EE (1978) Combined effects of testosterone and estradiol on rat ventral prostate in organ culture. Cancer Res 38:4126–4134

Fox LL, Redeuilh G, Baskevitch P, Baulieu EE, Richard-Foy H (1976) Production and detection of antibodies against the estrogen receptor from calf uterine cytosol. FEBS Lett 63:71–76

Garcia M, Rochefort H (1977) Androgens on the estrogen receptor. II. Nuclear translocation and uterine protein synthesis. Steroids 29:111–126

Gorski J, Toft DO, Shyamala G, Smith D, Notides A (1968) Hormone receptors: studies on the interaction of estrogen with the uterus. Recent Prog Horm Res 24:45–72

Gorski J, Sarff M, Clark J (1971) The regulation of uterine concentration of estrogen binding protein. Pergamon Press, Oxford (Advances in Biosciences, vol 7, pp 5–20)

Greene GL, Closs LE, Fleming H, de Sombre ER, Jensen EV (1977) Antibodies to estrogen receptor: imunochemical similarity of estrophilin from various mammalian species. Proc Natl Acad Sci USA 74:3681–3685

Higgins SJ, Rousseau GG, Baxter JD, Tomkins GM (1973) Early events in glucocorticoid action activation of steroid receptor and its subsequent specific nuclear binding studied in a cell free system. J Biol Chem 248:5866–5872

Horwitz KB, Costlow ME, MacGuire WL (1975) A human breast cancer cell line with estrogen, androgen, progesterone, and glucocorticoid receptors. Steroids 26:785–795

Hsueh AJ, Peck EJ, Clark JH (1975) Progesterone antagonism of the oestrogen receptor and oestrogen-induced uterine growth. Nature 254:337–339

Huggins C, Jensen EV (1955) The depression of estrone-induced uterine growth by phenolic estrogens with oxygenated functions at positions 6 or 16. The impeded estrogens. J Exp Med 102:335–346

Jensen EV, Jacobson HI (1962) Basic guides to the mechanism of estrogen action. Recent Prog Horm Res 18:387–414

Jensen EV, de Sombre ER (1972) Mechanism of action of the female sex hormones. Annu Rev Biochem 41:203–230

Jensen EV, Suzuki T, Kawashima T, Stumpf WE, Jungblut PW, de Sombre ER (1968) A two-step mechanism for the interaction of oestradiol with the rat uterus. Proc Natl Acad Sci USA 59:632–638

Jensen EV, Block GE, Smith S, Kyser K, de Sombre ER (1971) Estrogen receptors and breast cancer response to adrenalectomy prediction of response in cancer therapy. Natl Cancer Inst Monogr 34:55–70

Jost A (1961) The role of foetal hormones in prenatal development. Academic Press, New York (The Harvey lectures, vol 55, pp 201–226)

Jung-Testas I, Baulieu EE (1979) Effects of sex steroids and antihormones on growth adhesiveness and receptors of L–929 cells cultured in serum containing and serum free media. Exp Cell Res 119:75–85

Jung-Testas I, Desmond W, Baulieu EE (1976) Two sex steroid receptors in SC–115 mammary tumor cells. Exp Cell Res 97:219–232

Kahn CR (1976) Membrane receptors for hormones and neurotransmitters. J Cell Biol 70:261–281

Katzenellenbogen BS, Gorski J (1972) Estrogen action in vitro. Induction of the synthesis of a specific uterine protein. J Biol Chem 247:1299–1305

Kratochwil K, Schwartz P (1976) Tissue interaction in androgen response of embryonic mammary rudiment of mouse: identification of target tissue for testosterone. Proc Natl Acad Sci USA 73:4041–4044

Kuhn RW, Schrader WT, Smith RG, O'Malley BW (1975) Progesterone binding components of chick oviduct purification by affinity chromatography. J Biol Chem 250:4220–4228

Lebeau MC, Massol N, Baulieu EE (1973) An unsoluble receptor for oestrogens in the "residual" nuclear proteins of chick-liver. Eur J Biochem 36:294–300

Le Goascogne C, Baulieu EE (1977) Hormonally controlled "nuclear bodies" during the development of the prepuberal rat uterus. Biol Cell 30:195–206

MacGuire WL, Raynaud JP, Baulieu EE (eds) (1977) Progesterone receptors: Introduction and overview. Raven Press, New York (Progress in cancer research and therapy, progesterone receptors in normal and neoplastic tissues, vol 4, pp 1–8)

MacKnight GS (1978) The induction of ovalbumin and conalbumin mRNA by estrogen and progesterone in chick oviduct explant cultures. Cell 14:403–413

Massol N, Lebeau MC, Baulieu EE (1978) Estrogen receptor in hen oviduct chromatin, digested by micrococal nuclease. Nucleic Acids Res 5:723–738

Mercier-Bodard C, Alfsen A, Baulieu EE (1970) Sex steroid binding plasma protein. Acta Endocrinol [Suppl 147] (Kbh) 64:204–224

Mester J, Baulieu EE (1972) Nuclear estrogen receptor of chick liver. Biochim Biophys Acta 261:236–244

Mester J, Baulieu EE (1977) Progesterone receptors in the chick oviduct: determination of the total concentration of binding sites in the cytosol and nuclear fraction and effect of progesterone on their distribution. Eur J Biochem 72:405–414
Mester J, Martel D, Psychoyos A, Baulieu EE (1974) Hormonal control of oestrogen receptor in uterus and receptivity for ovoimplantation in the rat. Nature 250: 776–778
Mester J, Geynet C, Binart N, Baulieu EE (1977) Retarded administration of an antagonist stops steroid hormone action. Biochem Biophys Res Commun 79:112–118
Milgrom E, Baulieu EE (1969) A method for studying binding proteins, based upon differential dissociation of small ligand. Biochim Biophys Acta 194:602–605
Milgrom E, Atger M, Baulieu EE (1973a) Acidophilic activation of steroid hormone receptors. Biochemistry 12:5198–5205
Milgrom E, Luu Thi M, Atger M, Baulieu EE (1973b) Mechanisms regulating the concentration and the conformation of progesterone receptor(s) in the uterus. J Biol Chem 248:6366–6374
Mueller GC, Herranen AM, Jervell KF (1958) Studies on the mechanism of action of estrogens. Recent Prog Horm Res 14:95–140
O'Malley BW, Means AR (1974) Female steroid hormones and target cell nuclei. Science 183:610–620
Palmiter RD, Moore PB, Mulvihill ER, Emtage S (1976) A significant lag in the induction of ovalbumin messenger RNA by steroid hormones: a receptor translocation hypothesis. Cell 8:557–572
Palmiter RD, Mulvihill ER, MacKnight GS, Senear AW (1978) Regulation of gene expression in the chick oviduct by steroid hormones (Cold Spring Harbor Symposia on quantitative biology, vol 42 pp 639–647)
Radanyi C, Mercier-Bodard C, Secco-Millet C, Baulieu EE, Richard-Foy H (1977) Alpha-foetoprotein is not a subunit of the estradiol receptor of the rat uterus (rat alpha-foetoprotein/estrogen receptors). Proc Natl Acad Sci USA 74:2269–2272
Raynaud JP, Moguilewsky M (1976) Ontogenèse des récepteurs des oestrogènes chez le rat. In: Soulairac A, Gautray JP, Rousseau JP, Cohem J (eds) Système nerveux, activité sexuelle et reproduction. Masson, Paris, pp 85–92
Raynaud JP, Mercier-Bodard C, Baulieu EE (1971) Rat estradiol binding plasma protein. Steroids 18:767–788
Raynaud-Jammet C, Baulieu EE (1969) Action de l'oestradiol in vitro: augmentation de la biosynthèse d'ARN dans les noyaux utérins. C R Acad Sci [D] (Paris) 268:3211–3214
Renoir JM, Fox LL, Baulieu EE, Mercier-Bodard C (1977) An antiserum specific for human sex steroid binding plasma protein. FEBS Lett 75:83–88
Ringold GM, Yamamoto KR, Bishop JM, Varmus HE (1977) Glucocorticoid-stimulated accumulation of mouse mammary tumor virus RNA: increased rate of synthesis of viral RNA. Proc Natl Acad Sci USA 74:2879–2883
Robel P, Lévy C, Wolff JP, Nicolas JC, Baulieu EE (1978) Réponse à un anti-oestrogène comme critère d'hormono-sensibilité du cancer de l'endomètre. C R Acad. Sci [D] (Paris) 287:1353–1356
Rochefort H, Garcia M (1976) Androgens on the estrogen receptor. I. Binding and in vivo nuclear translocation. Steroids 28:549–560
Schimke RT, Pennequin P, Robins D, MacKnight GS (1977) In: Dumont J, Nunez J (eds) Hormones and cell regulation, vol 1. Hormonal regulation of egg-white protein. North Holland Publishing Company, Amsterdam, pp 209–221
Schwartz RJ, Kuhn RW, Buller RE, Schrader WT, O'Malley BW (1976) Progesterone-binding components of chick oviduct. In vitro effects of purified hormonereceptor complexes on the initiation of RNA synthesis in chromatin. J Biol Chem 251:5166–5177
Sica V, Nola E, Parikh I, Puca GA, Cuatrecasas P (1973) Purification of oestradiol receptors by affinity chromatography. Nature (New Biol) 244:36–39

Sutherland RL, Lebeau MC, Schmelck PH, Baulieu EE (1977a) Synergistic and antagonistic effects of progesterone and oestrogen on oestrogen receptor concentration and DNA polymerase activity in chick oviduct. FEBS Lett 79:253–257

Sutherland RL, Mester J, Baulieu EE (1977b) Hormonal regulation of sex steroid hormone receptor concentration and subcellular distribution in chick oviduct. In: Dumont J, Nunez J (Eds) Hormone and cell regulation, vol 1. North Holland publishing Company, Amsterdam, pp 31–48

Sutherland RL, Mester J, Baulieu EE (1977c) Tamoxifen is a potent "pure" antioestrogen in chick oviduct. Nature 267:434–435

Toft D, Gorski J (1966) A receptor molecule for estrogens: isolation from the rat uterus and preliminary characterization. Proc Natl Acad Sci USA 55:1574–1581

Toft D, Shyamala G, Gorski J (1967) A receptor molecule for estrogens: studies using a cell free system. Proc Natl Acad Sci USA 57:1740–1743

Truong H, Baulieu EE (1971) Interaction of uterus cytosol receptor with estradiol equilibrium and kinetic studies. Biochim Biophys Acta 237:167–172

Truong H, Geynet C, Millet C, Soulignac O, Bucourt R, Vignau M, Torelli V, Baulieu EE (1973) Purification of estradiol receptor by affinity chromatography. Representative experiments. FEBS Lett 35:289–294

Tseng L, Gurpide E (1975) Nuclear concentration of estradiol in superfused slices of human endometrium. J Clin Endocrinol Metab 41:402–404

Wade GN, Blaustein SD (1978) Effects of an antiestrogen on neural estradiol binding and on behaviors in female rats. Endocrinology 102:245–251

Wrange O, Gustafsson JA (1978) Separation of the hormone- and DNA-binding sites of the hepatic glucocorticoid receptor by means of proteolysis. J Biol Chem 253:856–865

Zava DT, MacGuire WL (1977) Estrogen receptor unoccupied sites in nuclei of a breast tumor cell line. J Biol Chem 252:3703–3708

Scanning Electron Microscopic Responses by the Endometrial Surface to Various Estrogens and Gestagens

B. O. Nilsson

Several scanning electron microscopic studies have demonstrated the marked changes that occur in the endometrial surface during the menstrual cycle (Nilsson and Nygren 1972; Ferenczy and Richart 1973; Hafez et al. 1975; Nathan et al. 1978; Ferenczy 1977). In short, the results of these studies show that a uterine surface in the follicular phase possesses slightly bulging cells with many long microvilli and that in the luteal phase it has cells with fewer and shorter microvilli but with comparatively large apical protrusions. Thus, in addition to available light microscopic signs already used in endometrial dating, it seems that the luminal surface changes also could be a useful indication on the type of hormone influencing the endometrium. Therefore, the present paper will review some recent results on hormonal responses by the endometrial surface of both menopausal and fertile women (Nathan et al. 1978; Nilsson et al., to be published; Englund et al., to be published).

The biopsies were obtained with the procedure common in clinical practice, and the specimens then were quickly fixed and prepared for scanning electron microscopy according to conventional critical-point-drying technique. In several cases, plasma values of hormones of interest have been determined.

Three main groups of hormonal treatments were tried on menopausal women: an estrogen treatment, a gestagen treatment, and a sequential one, i.e., at the end of an estrogen treatment a gestagen was added (Nathan et al. 1978; Nilsson et al., to be published). In each group, a few types of hormone and some different durations of treatments were tested (Tables 1 and 2). Further, the results of these experiments were exploited to evaluate the endometrial effects of hormones included in two types of contraceptive device: the subdermal implants and the vaginal ring (Englund et al., to be published) (Table 3).

The *control* patients, i.e., women in menopause, are not a homogeneous group since many of these women have a varying degree of persistant ovarian activity in addition to the estrogen production by the adrenal cortex. Analyses of some patients showed that the plasma estradiol -17β levels were 20–40 pg/ml, values about representative for women in menopause (Badawy et al. 1979). Compared to normal preovulatory levels of 200–300 pg/ml, these values are of course low but yet we do not know to what extent a target organ, like the uterine surface epithelium, responds to the endogeneous levels of estrogen. Therefore, some variation in the surface ultrastructure of the control patients has to be accepted.

The uterine epithelium of menopausal women had a surface ultrastructure indicative of a low cellular activity. Thus the epithelial cells were irregular in size and had a

Table 1. Type, dosage, and duration of hormone treatment of menopausal women

Case no.	Type and daily dose of hormone administered	Days of treatment
	Estrogen treatment	
1	Estradiol valerate 2 mg	1–20
2	Estradiol valerate 4 mg	1–17
3	Estradiol 4 mg + estriol 2 mg	1–14
4–6	Estriol 2 mg	1–14
7	Estriol 2 mg	1–28
8–10	Estriol 10 mg	1–14
	Gestagen treatment	
1–3	Progesterone 0.4 g	1–7
4–5	Progesterone 0.4 g	1–14

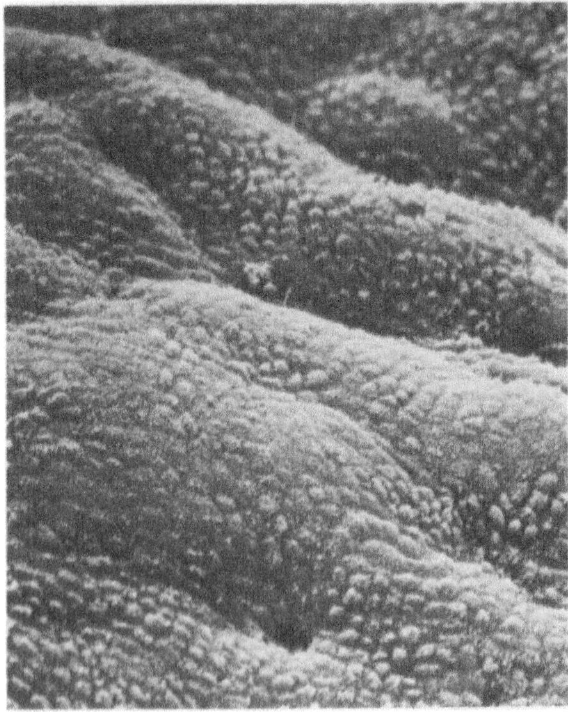

Fig. 1. Endometrium of menopausal woman. The luminal surface is undulating having gland-mouths situated in the pits. x 200

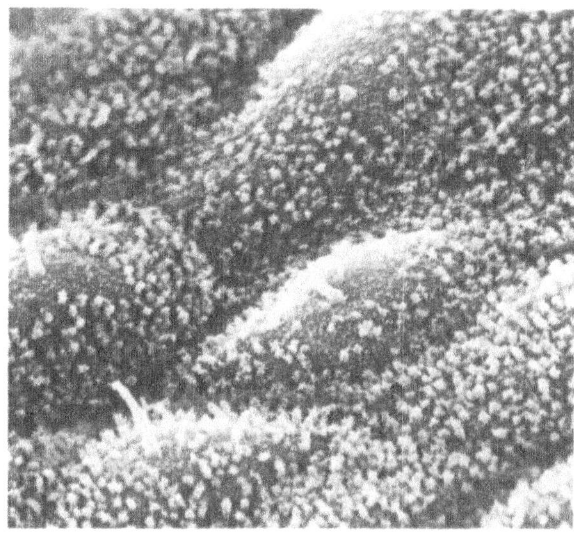

Fig. 2. Endometrium of menopausal woman. The epithelial cells are only slightly bulging and covered by a sparse amount of short microvilli. A single kinocilium is often noticed on the cells. This picture is representative for a postmenopausal epithelium. x 4,300

rather flat surface, often with only a few short microvilli (Fig. 1). A single kinocilium was a common finding (Fig. 2). Sometimes more microvilli were present, probably as a response to an increase of estrogens. Further, the surface morphology varied slightly locally, making it necessary to observe at least three to four specimens from each biopsy.

The *estrogen* treated women were given estradiol valerate, estradiol–17 β or estriol for between 2 and 4 weeks (Table 1). In the main, any estrogen treatment resulted in the appearance both of many ciliated cells, and of secretory cells with bul-

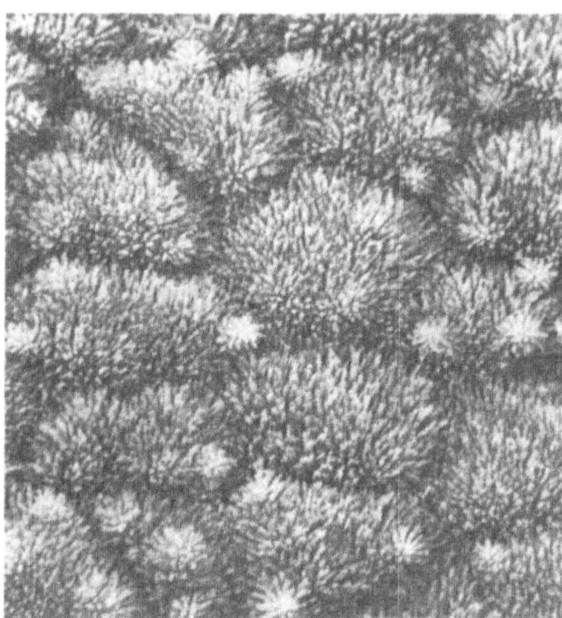

Fig. 3. Endometrium of menopausal woman treated with estradiol 4 mg and estriol 2 mg daily for 14 days. The bulging epithelial cells possess several long microvilli sometimes arranged in tuft-like groups. A growth of the microvilli is a typical response to estrogen. x 1,800

Fig. 4. Detail of endometrium demonstrated in Fig. 3 showing the long and slender microvilli. x 6,700

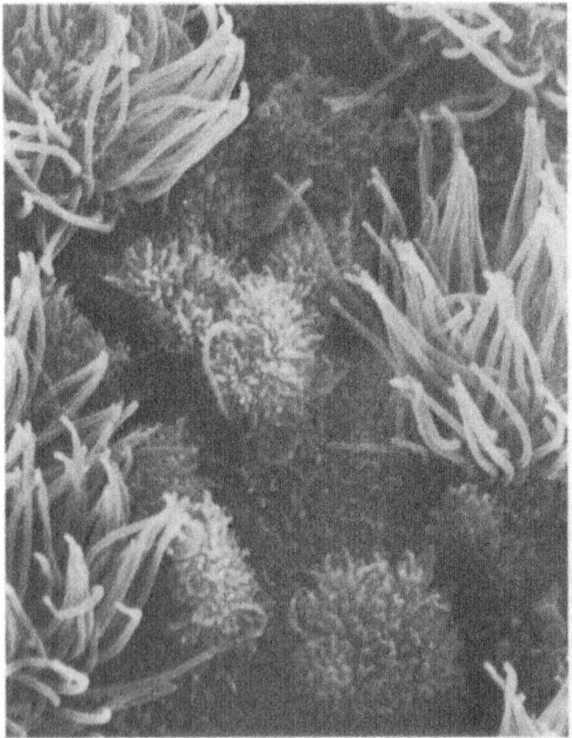

Fig. 5. Endometrium of menopausal woman treated with estriol 10 mg daily for 14 days. Bulging nonciliated cells are surrounded by some ciliated cells. The nonciliated cells display several long microvilli. x 3,600

ging surfaces and many long microvilli (Figs 3 and 4). It seemed that the more potent estrogen treatment given, the longer and more numerous were the microvilli. Tufts of microvilli, also appearing occasionally in a normal follicular phase, were another indication of an estrogen response by the cells (Figs. 3 and 4). Many of the specimens obtained from the estrogen-treated women also demonstrated a few small apical protrusions, but these were more common when in addition a gestagen had been administered.

Estriol given in a dose of 10 mg x 2 daily for about 2 weeks caused an epithelial response of the estrogen type: the number of ciliated cells increased and the secretory cells developed many long microvilli (Fig. 5). Although the general view is that estriol does not influence the endometrium, there are reports on responses by the uterine epithelium but only when administering daily doses of 6–8 mg for rather long periods (3–6 months) (Englund and Johansson, to be published). Evidently, the present technique – compared to the ordinary light microscopy mostly used in the previous studies – is a more sensitive one to trace minor responses by the uterine epithelium.

A *gestagen* treatment (Table 1), in form of suppositories of progesterone 0.2 g x 2 daily for up to 14 days, raised the plasma levels from less than 1 ug/liter to about 6 ug/liter. The morphological result, however, was meager, and no marked signs of any specific changes could be found. Also, there is no report as yet that progesterone can transform an endometrium, earlier not having been subjected to an estrogen treatment.

Table 2. Type, dosage, and duration of hormone treatment of menopausal women

Case no.	Type and daily dose of hormone administered	Days of treatment
	Sequential treatment	
1–2	Estradiol 10 mg	1–21
	Progesterone 0.4 g	15–21
3–4	Estradiol valerate 2 mg	1–19
	Norgestrel 0.5 mg	12–19
5–6	Estradiol valerate 2 mg	1–21
	Progesterone 0.4	15–21
7	Estradiol valerate 2 mg	1–22
	Noretisterone 5 mg	15–22
8	Estradiol 2 mg-estriol 1 mg	1–22
	Noretisterone 5 mg	15–22
9	Estradiol 2 mg-estriol 1 mg	1–21
	Progesterone 0.2 mg	15–21
10–12	Estriol 10 mg	1–21
	Progesterone 0.4 mg	15–21

The *sequential* treatment tested included various estrogens and gestagens (Table 2). As expected, the treatment resulted in a surface ultrastructure similar to what characterizes a normal premenstrual uterine surface: bulging cell surfaces with microvilli and large morel-like apical protrusions (Figs. 6 and 7). Further, even a priming with estriol was effective and resulted in a moderate development of the apical protrusions.

Fig. 6. Endometrium of menopausal woman treated with estradiol valerate 2 mg daily for 19 days and with norgestrel 0.5 mg daily for the last 8 days. Among the tips of kinocilia, apical protrusions are morel-like having a present. The apical protrusions are morel-like having a wrinkled surface. These protrusions are signs of a secretory transformation of the endometrium caused by the addition of a gestagen. x 5,400

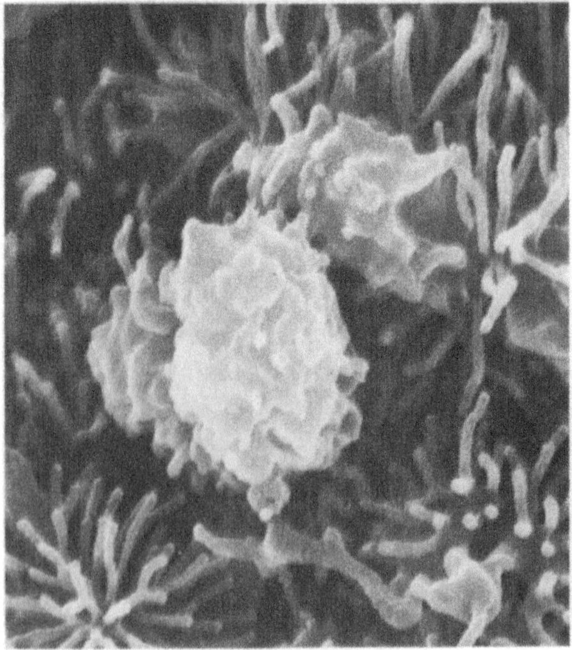

Fig. 7. Endometrium of a menopausal woman treated with estradiol 2 mg and estriol 1 mg daily for 21 days and with progesterone 0.2 mg two times a day for the last 7 days. Detail of nonciliated cells demonstrating the ruffled apical protrusions and the long and slender, and sometimes branched, microvilli. x 6,700

Table 3. Type of contraceptive device used and type of hormone included

Fertile women carrying
1. Subdermal implants with norgestrel
2. Intravaginal rings with norgestrel
3. Intravaginal rings with norgestrel and estradiol
Menopausal women carrying
4. Intravaginal rings with estradiol

The length of the microvilli was influenced by the estrogen given: short microvilli were associated with estradiol valerate and long microvilli with estradiol. Also the size of the apical protrusions changed with the hormone administered: small protrusions after the 19-nor-testosterone derivates but large protrusions after progesterone.

The influence of two types of *hormone releasing contraceptive device* on the uterine surface of fertile women was examined, namely subdermal implants and vaginal rings (Englund et al., to be published) (Table 3). The implants released L-norgestrel, giving a norgestrel concentration in plasma between 0.5 and 1 ng/ml. The estradiol levels in plasma were below 100 pg/ml, that is, in the range usually found during early follicular phase of the menstrual cycle. When the biopsies were taken, the implants had been in place for about a year. Also the vaginal rings contained L-norgestrel, and the release gave rather similar plasma levels of norgestrel and estradiol as did the subdermal implants. These biopsies were taken after an 18-day treatment.

SEM demonstrated cells with rather flat surfaces and short microvilli (Fig. 8). This appearance could be called an inactive one and is in accordance with what could be implied from light microscopic reports on the influence of gestagens on a cycling endometrium (Nilsson et al. 1978; Martinez-Manautou et al. 1975; Scommegna et al.

Fig. 8. Endometrium of fertile woman carrying an intravaginal ring releasing norgestrel. Ciliated cells are intermingled among rather flat nonciliated cells possessing a few short microvilli. This picture indicates an inactive luminal epithelium. x 1,800

Fig. 9. Endometrium of fertile women carrying an intravaginal ring releasing norgestrel and estradiol. The nonciliated cells are bulging being covered by several long microvilli, i.e., demonstrating an estrogen response. x 2,500

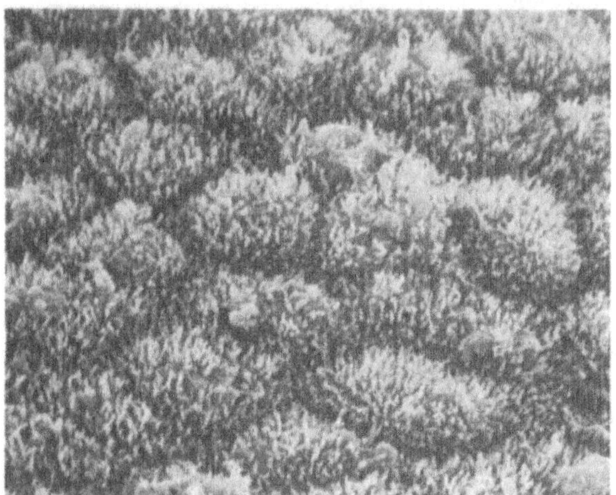

Fig. 10. Endometrium of menopausal woman carrying an intravaginal ring releasing estradiol. Bulging microvillous cells being representative for an estrogen response. x 1,800

1970). Now, it has been shown that adding estradiol to the norgestrel load of a vaginal ring improves the bleeding pattern. Using rings that release approximately 200 μg estradiol per day results in blood plasma levels of 25–50 pg/ml. Biopsies taken a month after the insertion of the device revealed changes of the uterine surface structure: the cells were covered with many long microvilli, i.e., an estrogen response had appeared (Fig. 9). A similar response was also observed in the endometrium of menopausal women carrying an estradiol-containing vaginal ring for 3 weeks (Fig. 10). Pretreatment plasma estrogens were in the postmenopausal range, while treatment levels of estradiol varied between 100 and 200 pg/ml plasma. Thus, the vaginal administration of estrogen is an effective way to influence the endometrial structure.

In conclusion, scanning electron microscopy of endometrial biopsies is a useful tool when evaluating the influence of various hormones on the uterine surface epithelium. But interpretations of the results are still hampered by some lack of knowledge. For instance, neither do we known the sensitivity in the responses by the target cells to changes in the hormonal plasma levels, nor do we know how long a certain structural change remains in the target cells after vanishing of the eliciting hormone from the blood plasma.

References

Badawy SZA, Elliot LJ, Elbadawy A (1979) Plasma levels of oestrone and oestradiol -17β in postmenopausal women. Br J Obstet Gynaecol 86:56–63

Englund D, Johansson EDB (to be published) Endometrial effect of oral oestriol treatment in postmenopausal women. Acta Obstet Gynecol Scand

Englund D, Nilsson BO, Weiner E, Victor A (to be published) Endometrial effects of levonorgestrel and estradiol. A scanning electron microscopical study of the luminal epithelium. Contraception

Ferenczy A (1977) Surface ultrastructural response of the human uterine lining epithelium to hormonal environment. A scanning electron microscopic study. Acta Cytol (Baltimore) 21:566–572

Ferenczy A, Richart RM (1973) Scanning and transmission electron microscopy of the human endometrial surface epithelium. J Clin Endocrinol Metab 36:999–1008

Hafez ESE, Ludwig H, Metzger H (1975) Human endometrial fluid kinetics as observed by scanning electron microscopy. Am J Obstet Gynecol 122:929–938

Martinez-Manautou J, Maqueo M, Aznar R, Pharriss BB, Zaffaroni A (1975) Endometrial morphology in women exposed to uterine systems releasing progesterone. Am J Obstet Gynecol 121:175–179

Nathan E, Knoth M, Nilsson BO (1978) Scanning electron microscopy of the effect of short-term hormonal therapy on postmenopausal endometrium. Ups J Med Sci 83:175–183

Nilsson O, Nygren KG (1972) Scanning electron microscopy of human endometrium. Ups J Med Sci 77:3–7

Nilsson CG, Luukkainen T, Arko H (1978) Endometrial morphology of women using a D-norgestrel-releasing intrauterine device. Fertil Steril 29:397–401

Nilsson BO, Nathan E, Knoth M (to be published) Scanning electron microscopical responses by postmenopausal endometrium on treatments with oestriol and oestriol-progesterone. Ups J Med Sci

Scommegna A, Pandya GN, Christ M, Lee A, Cohen MR (1970) Intrauterine administration of progesterone by a slow releasing device. Fertil Steril 21:201–210

Morphological Studies of Structural Changes in the Tubal Mucosa of the Rabbit at Estrus and During HCG-Induced Pseudopregnancy

W. Kühnel

Introduction

The first anatomic description of the oviduct was given by Gabriele Fallopius in 1561 (Woodruff and Pauerstein 1969). Its function, however, remained a matter of conjecture for a very long time. Today we have advanced to the point to postulate that the oviduct does more than merely transport the egg from the ovary to the uterus by the motion of its mucosa and muscle layer; we know that it is important to the development of the ova from the time it is fertilized in the ampulla until it is embedded in the uterus. Nevertheless, our knowledge is still incomplete about the precise relation between tubal mucosa and ova-to-blastocyst development (Pauerstein and Eddy 1979).

The ovarian hormones control both morphological and biochemical changes that the mucosa of tuba uterina and uterus undergo during all phases of reproduction. Following ovulation these transformations take on a functional meaning, for at this point the genitalia must adapt to the requirements of the developing embryo (Beier and Kühnel 1976).

The most important stages of early pregnancy – fertilization of the ova, its passage through the tube and development into a blastocyst, and finally its implantation – can all be observed with particular clarity in the rabbit. This laboratory animal has a provoked ovulation, which is physiologically induced by copulation and the ensuing secretion of gonadotropin by the pituitary; ovulation occurs about 10.5 h post coitum. The ova enter the fallopian tube via the ostium tubae and begin their passage, which takes about 70 h. In the laboratory one can stimulate ovulation by injecting gonadotropin (20–50 IE HCG i v); natural or artificial insemination will then lead to fertilization and normal pregnancy. If the ova are not fertilized, however, pseudopregnancy will ensue (Beier 1968; Kühnel et al. 1971; Beier and Kühnel 1973).

In the rabbit, ovulation triggers the uterine cycle and estrus gives way to the changes of the preimplantation phase. To form a picture of how the maternal organism adapts structurally to the embryonal system during the period from copulation to implantation, we performed morphological investigations of the secretion of the endosalpinx epithelium. The peripheral plasma concentrations of the ovarian hormones progesterone and 17β-estradiol are very high immediately after copulation. Both hormones show a marked decrease on the lst and 2nd days post coitum; on day 3 ovarian hormone secretion begins to return to former levels (Beier and Kühnel 1976; Challis et al. 1973; Fuchs et al. 1974; Hasan et al. 1971; Hilliard and Eaton 1971).

Surface Relief of Mucosa

The descriptions we have of the surface relief of tubal mucosa are all very vague. Horstmann and Stegner (1966) tell us that the folds in the pars ampullaris of the human oviduct are "extraordinarily dense and fill the entire space except for a labyrinth of narrow channels." These authors admit that "the longitudinally opened duct gives only a vague impression of the complicated folds of the pars ampullaris, since the secondary and tertiary folds are usually stuck together." They go on to describe how the folds appear to be lower and closer together in the constricted parts of the oviduct, and say that in the isthmus, apart from three to five ridgelike protuberances, there is no mucosal folding at all.

No really precise information is available on the surface relief of the rabbit oviduct, despite many reports that have been published on the structure of oviductal epithelium (Nilsson and Reinius 1969; Greenwald 1969; Beck and Boots 1974). Only Hafez and Kanagawa (1975) tell us that "... the mucosal folds of the fimbriae are arranged in an irregular pattern" and "... in general, the oviductal mucosal folds are arranged longitudinally from the isthmus to the ampulla."

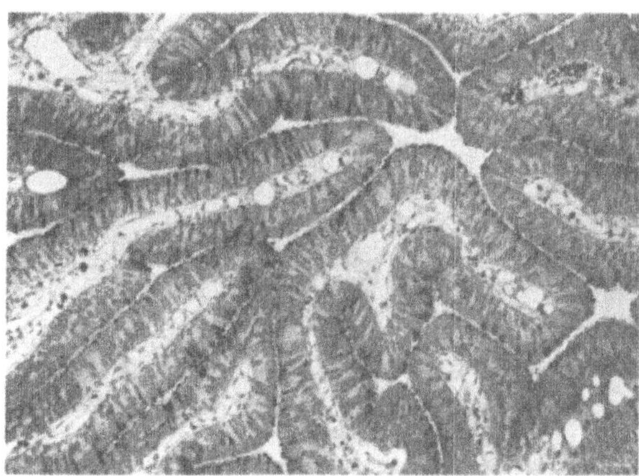

Fig. 1. Cross-section of pars ampullaris tubae showing mucosal folds which fill the entire space except for narrow gaps. Semithin section, Richardson stain. x 93

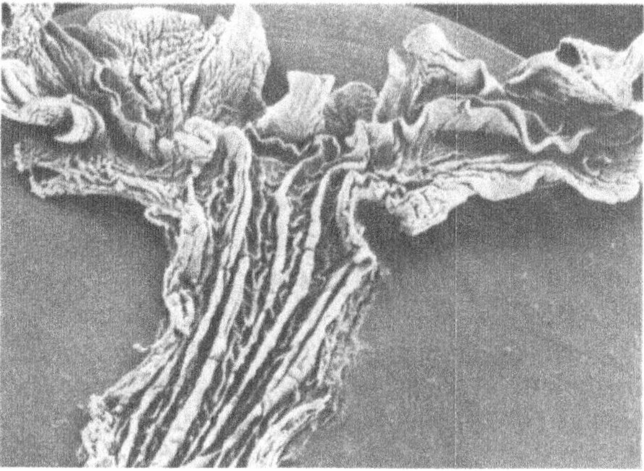

Fig. 2. Mucosal relief on longitudinal section of the oviduct showing lobe-shaped oviductal fringe. Scanning electron micrograph. x 6.67

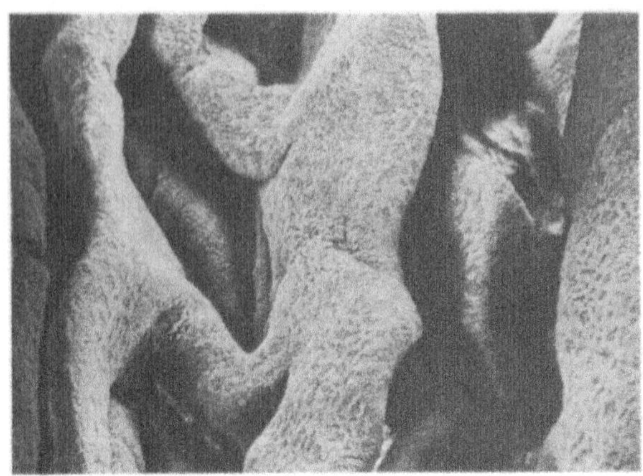

Fig. 3. Mucosal protuberances from the pars ampullaris tubae. Scanning electron micrograph. x 50

The rabbit oviduct is divided into preampulla with infundibulum, ampulla, isthmus, and junctura (Nilsson and Reinius 1969). On sections, each of these areas reveals a characteristic mucosal formation. In the ampulla we see extremely long, thin villi which are often bent or folded (Fig. 1); in the isthmus, broad-based, leaf-shaped protuberances which are rich in connective tissue and only very rarely branched. Seen under a scanning electron microscope, however, the oviductal mucosa of rabbits in estrus shows a type of relief that these sections do not reveal.

Mucosal ridges emerge from the lobe-shaped fimbriae to form longitudinal folds, between which, however, we can detect a further system of fine corruscations (Fig. 2). These are intricately branched; some of the branches dissipate, others are anastomosed with one another (Fig. 3). The modeling of the surface often runs across or diagonal to the longitudinal axis of the oviduct. Between these branched ridges lie flat depressions, deep valleys, and narrow channels.

In the isthmus, these longitudinal mucosal folds are even more developed; they are thicker and more extended. Deep in the canyons between them appear narrow bridges that connect the longitudinal folds with one another laterally. Having seen these pictures it is difficult to go on speaking of the oviduct as having only longitudinally arranged, regular mucosal folds; or in other words, our spatial conceptions of the form and extent of the mucosal relief in the rabbit oviduct do not conform to the true picture. For example, the preampulla is not covered with fimbriae as we had assumed; its funnel-shaped first section is divided into two to three lobelike mucosal plates that merge gradually into the wall of the tube. The long mucosal villi that have been described on the basis of sections do indeed represent the relatively high ridges and folds that run longitudinally down the oviduct; these, however, are not branched at all, so that we cannot speak of them as secondary or tertiary folds in the usual sense. The complicated system of low ridges in the valleys between these large mucosal folds has nothing to do with the intricately branched trees that have been described on the basis of sections. Rather, they are folds in their own right which connect, laterally or diagonally, the higher, longitudinally oriented mucosal folds with one another (Busch et al. 1977; Kühnel and Busch 1979). This mucosal modeling does not change during the preimplantation phase.

Oviductal Epithelium During Estrus

In the highly prismatic epithelial layer of the oviduct we find almost equal numbers of secretory and ciliated cells arranged in a relatively regular pattern – at least this is true of the ampulla (Fig. 4). In the isthmus, secretory cells clearly predominate. On semithin sections stained by the Richardson method the clusters of secretion granules at the cell apices appear very clearly, as do the slightly domed cells themselves (Fig. 5). This finding is confirmed by scanning electron microscopy. Here we see, protruding from between the ciliated cells, large numbers of secretory cells covered with short, stumplike microvilli (Fig. 6). The surface of the infundibulum is made up of ciliated cells almost exclusively (Fig. 7). Here the cilia appear to be arranged in rows, between which slender microvilli protrude at intervals. At the ampulla end of the oviduct the secretory cells carry spherical or buttonlike formations of various sizes on their apices (Fig. 8).

We can add to what was said in an earlier work on the oviductal epithelium (Brenner 1969): the Richardson stain brings both ciliated and secretory cells into unsually high relief. Note, however, that the cytoplasm does not stain uniformly – secretory cells are deeper in color than ciliated cells (Fig. 4). Also, the position of the nucleus is

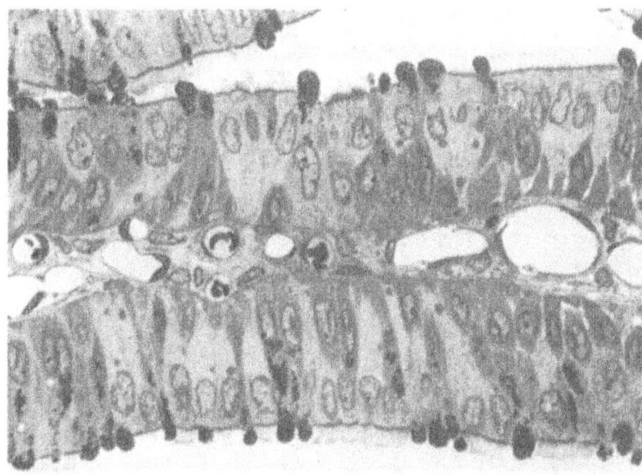

Fig. 4. Mucosal fold from the pars ampullaris tubae. Note epithelial tissue with light-colored ciliary cells and darker secretory cells, which have protruding apices. These protrusions are filled with secretion granules. Semithin section, Richardson stain. x 373.5

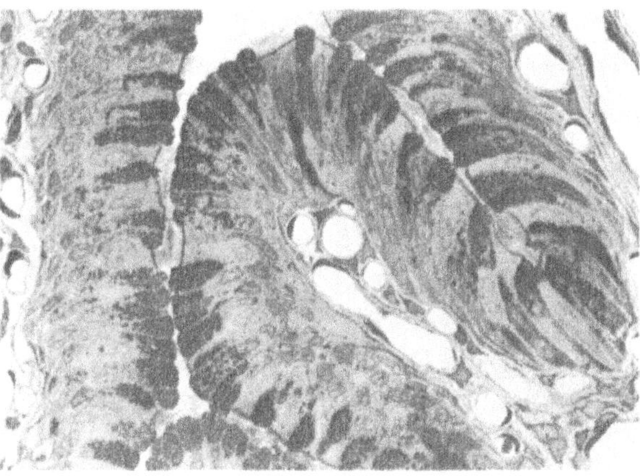

Fig. 5. Mucosal folds from pars isthmica tubae showing epithelium with large numbers of granule-containing secretory cells, and isolated ciliary cells. Semithin section, Richardson stain. x 373.5

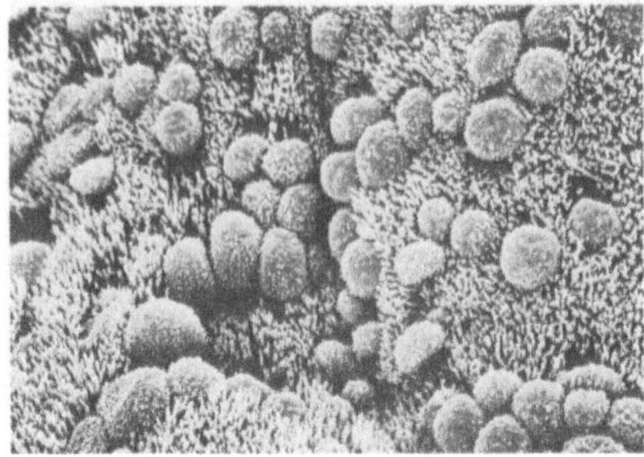

Fig. 6. Top view of a mucosal fold from the pars ampullaris tubae, showing numerous ciliary cells between which secretory cells protrude. Scanning electron micrograph. x 1,152

characteristic for each type of cell; in ciliated cells it is located in the upper third and in secretory cells in a basal position. The cytoplasm above the nucleus is filled with intensively stained granules (Figs. 4, 5). The protruding cell apices also contain densely packed secretion granules (Fig. 4), which appear even more clearly on semithin sections of the isthmus stained with toluidin blue (Fig. 5).

Histochemically, the granular regions of the secretory cells react strongly with alcian blue stain at pH 0.5 and pH 2.5; the PAS reaction is positive and is not weakened if the cells are digested beforehand by diastasis (Figs. 9a, 10a). This probably means that the granules are composed of neutral and acidic mucoproteins (cf. Denker 1970 a, b).

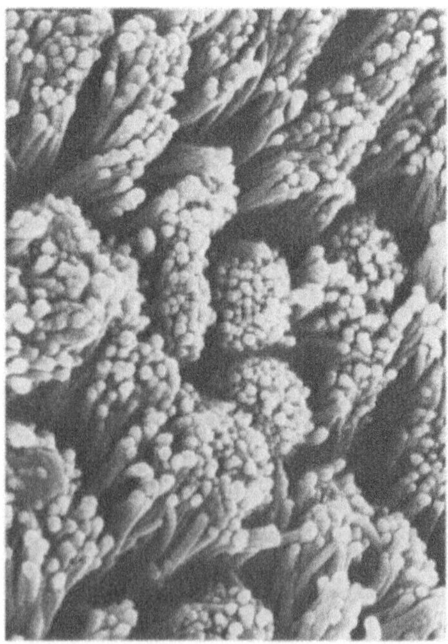

Fig. 7. Bunched cilia in the infundibular region. Scanning electron micrograph. x 2,832

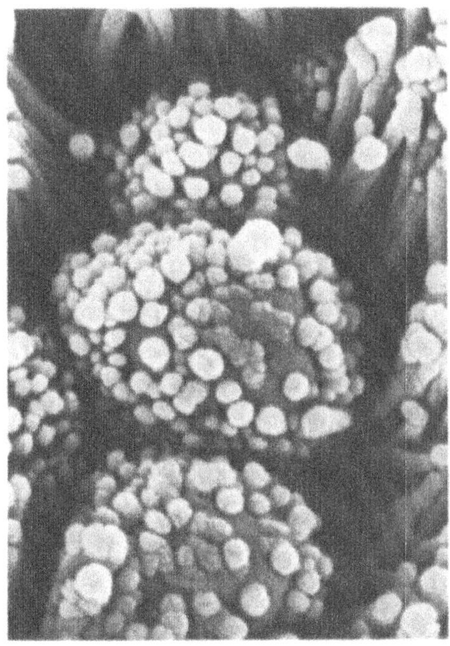

Fig. 8. Pars ampullaris tubae. Note the three secretory cells whose tops are covered with spherical projections. Scanning electron micrograph. x 5,664

Field electron micrography of the oviductal mucosa shows that the epithelial tissue consists of the following types of cell – secretory, ciliated, and granulated basal (Fig. 11). We will discuss the fine structure of each of these cell types below, emphasizing the findings obtained by higher magnification.

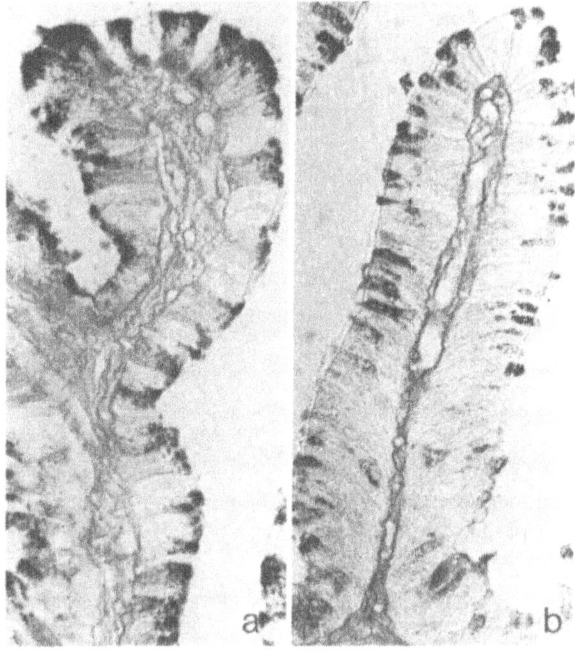

Fig. 9. Pars ampullaris tubae. at estrus, **b** on day 6 pc. Alcian blue stain at pH 2.5. Note the reduction in alcianophilic substances on day 6 pc. x 268.8

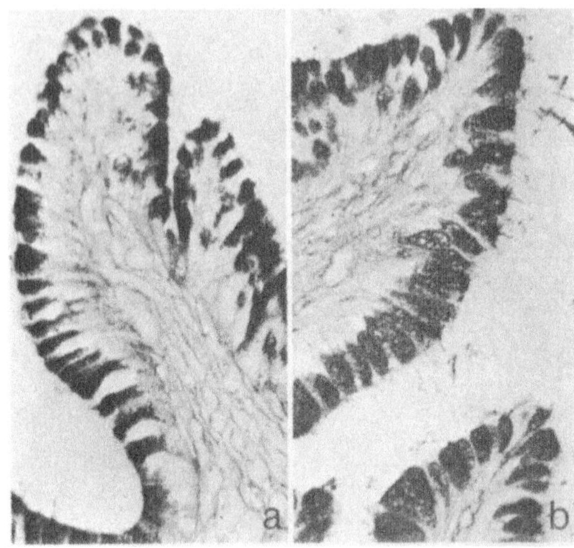

Fig. 10. Pars isthmica tubae. **a** at estrus, **b** on day 6 pc. Alcian blue stain at pH 2.5. On day 6 pc the alcianophilic substances are less dense. x 298.9

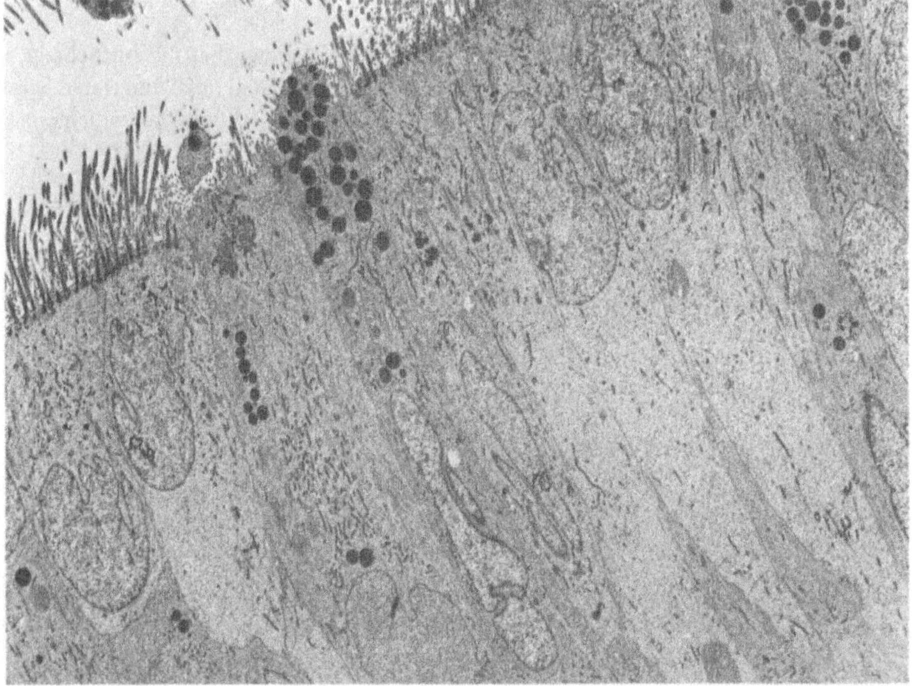

Fig. 11. Vertical section through the oviductal epithelium of the pars ampullaris. The ciliary and secretory cells have osmiophilic granules; the former a light cytoplasm and nuclei near the apex. Electron micrograph. x 2,171

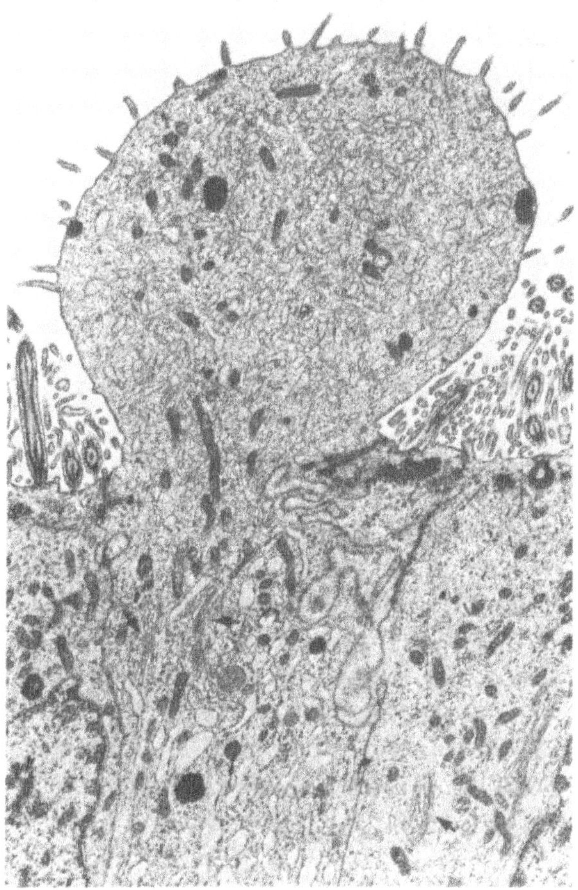

Fig. 12. Pars ampullaris tubae, showing a secretory cell sandwiched between two ciliary cells. Note the balloonlike protrusions with mitochondria and numbers of rER tubules, isolated microtubuli and secretion granules. Golgi apparati are indicated by an *arrow*. Electron micrograph. × 8,895

Secretory Cells

The characteristic secretory cells in the ampulla are elongated, slender or bellied, and have apices that protrude far into the lumen. As a rule they are covered on all sides by plasmalemma (Figs. 12, 13, 14). Their protruding apices are extremely diverse in shape. Next to gentle domes one finds large, balloonlike configurations (Fig. 12), pear-shaped cells, and cells with stalks that resemble bottlenecks (Figs. 13, 14). Some of these cell protrusions are covered with short, stumpy microvilli. In their finely granulated cytoplasm we find osmiophilic secretion granules of various density, size, and shape, and also slender mitochondria of the crista type, microtubuli arranged either singly or in groups, some filaments, and lysosomelike particles. Some of the ballooned out cells are filled with large numbers of tiny bubbles and tubular structures, among which we detect small mitochondria and isolated osmiophilic secretion particles (Fig. 12). We believe that these bubbles and tubules represent sections of endoplasmic reticulum whose membranes have only a few remaining ribosomes and whose spaces are filled with fine flakes of weakly osmiophilic material.

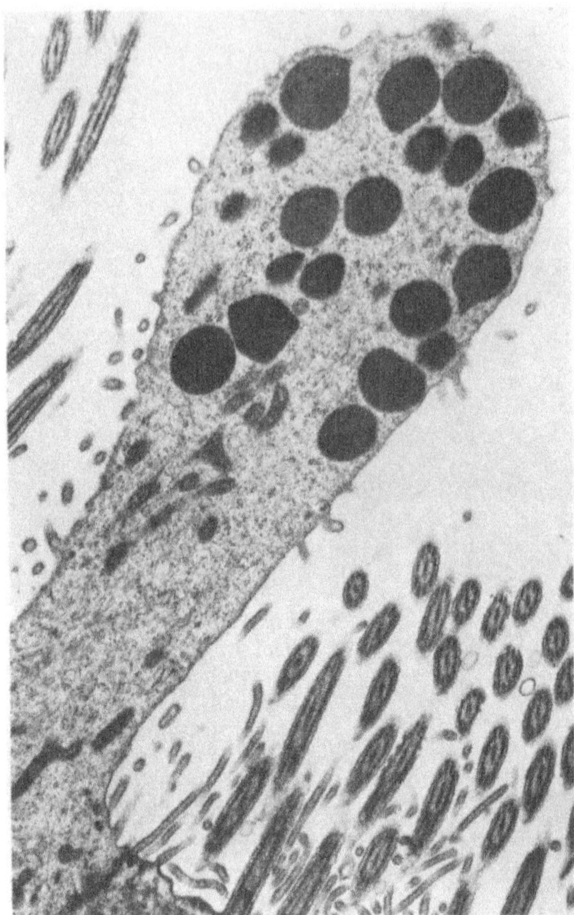

Fig. 13. Pars ampullaris tubae, showing secretory cell protruding far into the oviductal space. The cytoplasm of this protrusion contains osmiophilic secretion granules, mitochondria, and isolated microtubuli. Electron micrograph. x 16,584

In discussing our light-microscopic findings, we will give special attention to the basal, basolateral, and supranuclear regions of the secretory cells. The ground plasma located directly under the apex of these cells is often more densely structured than that deeper down in the cell body and contains osmiophilic secretion granules. The supranuclear and basolateral regions are characterized by high numbers of mostly extended mitochondria which have markedly dense matrices, and also by well-developed rough endoplasmic reticulum which is often distended into cisterns. The golgi apparatus, located supranuclearly or often apically, is generally well developed. It consists of stacked membranes, and small or ligatured vacuoles which contain electron-pervious material. In the vicinity of the golgi apparatus we often find secretion granules of varying density (Fig. 15). Sometimes thin lines of microtubuli are detected, and in many areas of the cell, lysosomes, free ribosomes in groups, and filaments. The nuclei, which are flattened oval in shape, lie in a basal position; they are often dented or notched; and the chromatin is frequently found in high concentrations forming particles at the nuclear membrane.

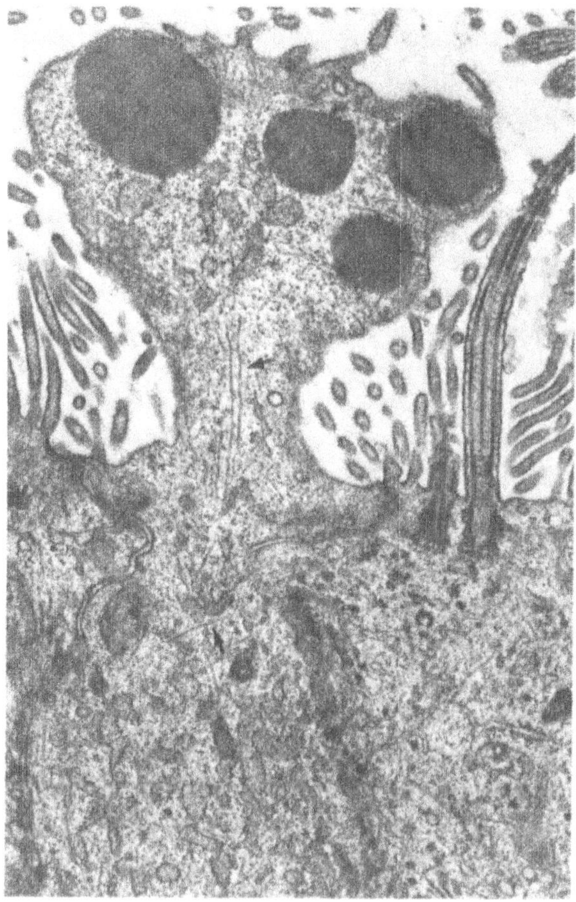

Fig. 14. Pars ampullaris tubae, showing secretory cell between two ciliated cells. Note thin-necked protrusion. Microtubuli are indicated by an *arrow*. Electron micrograph. x 17,640

Though the secretory cells all abut on the basal lamina (which is osmiophilic), they often reach it only by slender projections of the most varied shape; in consequence, the secretory cell base is rough and uneven, leaving large, empty regions that often extend in a basolateral direction. Usually, however, the intercellular spaces are narrow and distended, and are sealed off just before they reach the surface by a network of terminal bars and desmosomes.

Electron micrography of the epithelium in the constricted sections of the oviduct reveals that the secretory cells here take on a marked club shape, tapering in the basal direction, with their narrow bases butting on the basal lamina. The nucleus, which is flatly oval in shape and notched, is located in the narrow part of these cells, as are the organelles typical of secretorily active cells. Particularly noteworthy are the very prominent golgi apparatuses in the supranuclear cytoplasm (Fig. 16).

Our electron micrographs confirm our suspicion that it is incorrect to think of the secretion granules as being uniform. This secretory material is actually much more complexly structured than it appears under the light microscope. The granules are heteromorphic both in shape and size, and also in terms of their matrix structure; it is remarkable how many different types are found side by side in one cell (Fig. 17). On

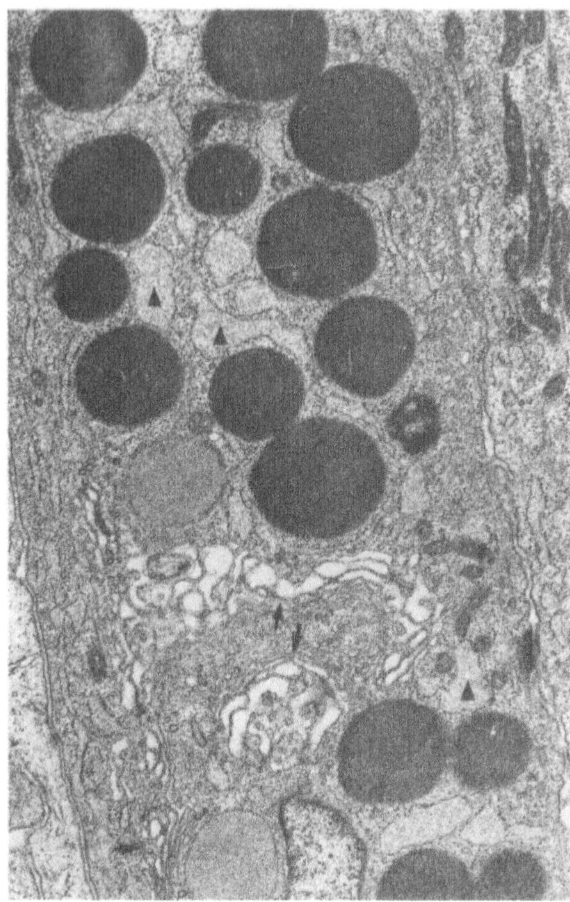

Fig. 15. Pars ampullaris tubae. Section of a secretory cell, showing supranuclear region of the cytoplasm with golgi field (↑), secretion granules and cisterns of rER (▲). Electron micrograph. x 16,464

overview pictures spotted granules are obvious; they are tightly packed and separated only by narrow peninsulas of cytoplasm. In other cases the membrane is missing, allowing neighboring secretion granules to merge. Their content is very light in color and consists of fine, dustlike material. Numerous small, osmiophilic grains with indeterminate borders are seen, and in some cases one or several larger, sickle-shaped dense regions that are usually located eccentrically near the periphery of the granular accumulation. These regions also contain of very fine particles that are impervious to electrons. In other secretion granules we find a thin, slightly wrinkled membrane enclosing dustlike material of varying osmiophilia. The mode of secretion of these granules during estrus has yet to be convincingly explained. We were able to find no ruptured membranes at the cell apex during this stage – openings through which secretion could enter the oviduct. Thus we cannot confirm the findings of Brower and Anderson (1969), who reported that secretion was extruded.

As far as other characteristics are concerned, the secretory cells in the isthmus area never have the impressively protruding apex that is typical of those in the ampulla. They jut only a short distance into the oviduct; sometimes their apex carries short microvilli. The cell sides are almost straight and run parallel down to the basal third.

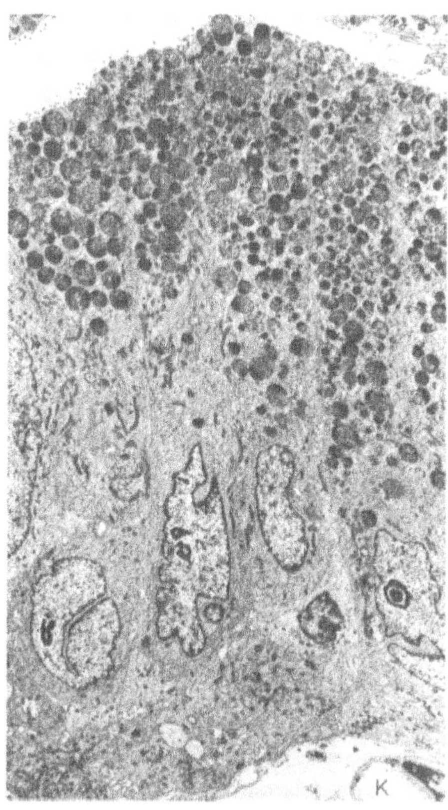

Fig. 16. Pars isthmica tubae, showing several secretory cells with numbers of heteromorphic secretion granules. Cell nuclei in a basal position. Note subepithelial connective tissue with capillaries *(K)*. Electron micrograph. x 2,742

At this point one often sees interdigitating plasmalemmata, and farther down toward the base of the cell, the intercellular spaces grow wider. Contact between the upper third of these cells is by means of zonulae occludentes and desmosomes.

At the base, in the niches between neighboring secretory and ciliated cells, we have observed markedly dense elements that are often triangular in shape and abut on the basal lamina with one of their broad sides (Fig. 18). These are doubtless the so-called granulated basal cells mentioned above. However, under the electron microscope these "granules" appear to be small, usually sausage-shaped mitochondria with an extremely dense matrix. As far as we can tell, mitochondria of the tubular type may also occur. Thus it is still an open question whether what we are seeing is a special third kind of cell or whether these elements might be thought of as replacements for exhausted secretory cells.

Ciliated Cells

The ciliated cells in the oviductal mucosal epithelium (Fig. 19) are highly prismatic and bound to neighboring cells by terminal bars and desmosomes. The ciliary structure is no different in principle from that of kinocilia (Bargmann 1977; Fawcett and Porter 1954). The plasmalemma at the cell apex protrudes from between the cilia as simple microvilli, some of which are short, some long or branched (cf. Horstmann and Steg-

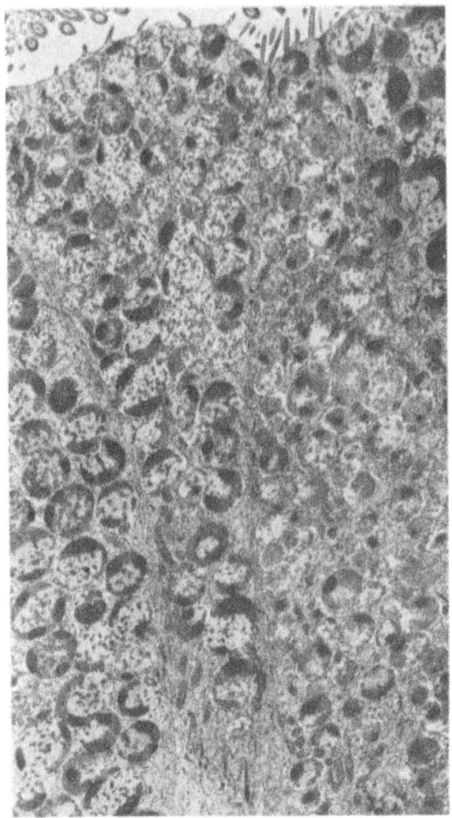

Fig. 17. Pars isthmica tubae, showing four secretory cells, their supranuclear regions filled with heteromorphic secretion granules. Electron micrograph. x 5,484

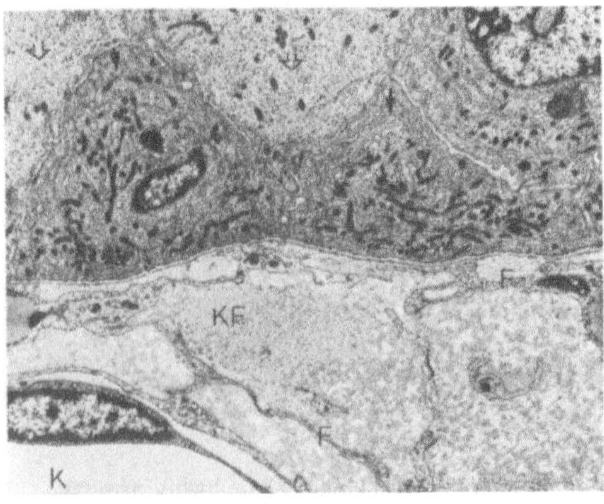

Fig. 18. Pars isthmica tubae, showing granulated basal cells (↗), basal sections of secretory cells (⇧), subepithelial connective tissue with capillaries *(K)*, collagen fibers *(KF)* and fibrocyte appendages *(F)*. Electron micrograph. x 9,810

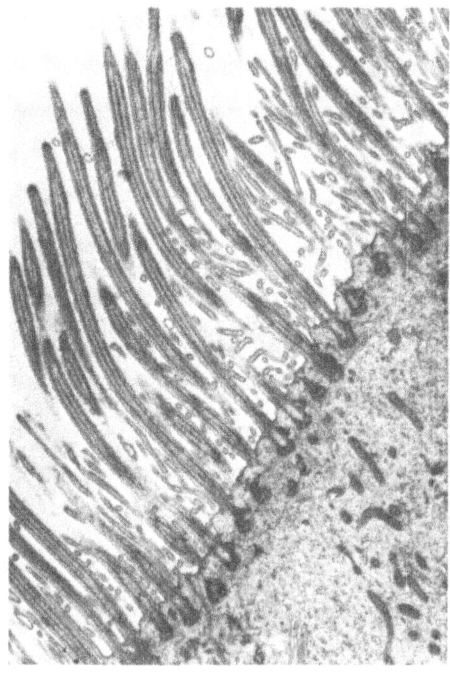

Fig. 19. Pars ampullaris tubae. Ciliary cells with cilia and microvilli. Electron micrograph. x 10,200

ner 1966). As already mentioned above, there is a marked difference in density between ciliated and secretory cells, even under the light microscope. The very light cytoplasm of the former also shows up under the electron microscope. The ground cyto plasm of ciliated cells is very permeable to electrons; it contains extremely fine grains of varying density and isolated, slender mitochondria of the crista type. The nucleus, an extended oval with many notches, is found near the cell apex. The cytoplasm is denser in the perinuclear region than in the infranuclear and basal regions. It usually contains several golgi fields, a considerable number of club-shaped or folded mitochondria with dense matrices, isolated multivesicular bodies and lysosomes, and electron-impermeable granules which possibly represent secretion granules. Just below the row of basal bodies the cytoplasm is often shot through with microtubuli and extremely fine filaments, which are sometimes bunched together. Also of note is the fact that glycogen deposits are found in almost all ciliated cells. These are limited agglomerations of glycogen and lie in the basal cytoplasm (cf. Horstmann and Stegner 1966).

Oviductal Epithelium in the Preimplantation Phase
[Day 0 to Day 6 post coitum (p.c.)]

In the following we shall discuss the changes that occur in oviductal epithelium during the preimplantation phase of pregnancy. Even under the light microscope, samples of the ampulla stained by the Richardson method (araldite sections with toluidin blue) reveal that the secretory cells have less pronounced apices than before, and that they contain fewer granula (Fig. 20).

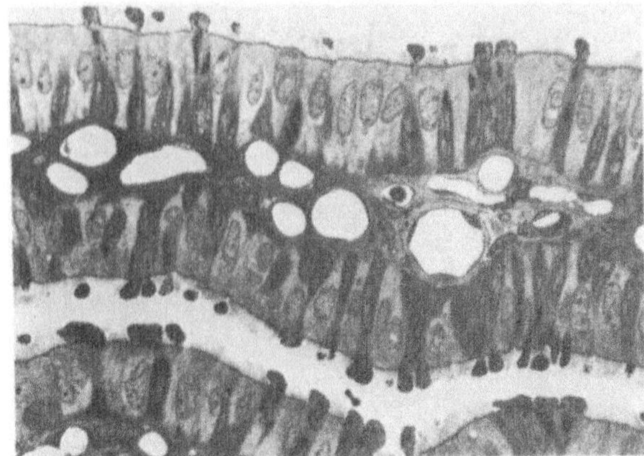

Fig. 20. Mucosal fold in pars ampullaris tubae on day 6 pc, showing epithelial tissue with light ciliated cells and dark secretory cells. Note that the protrusions have only a low granule content, cf. Fig. 4. Semithin section, Richardson stain. x 373.5

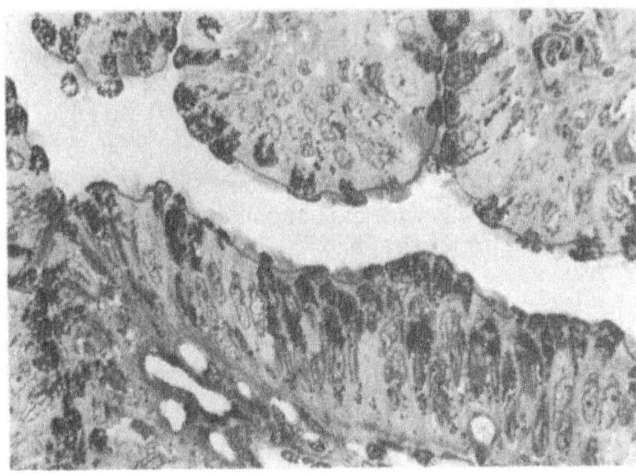

Fig. 21. Mucosal folds from pars isthmica tubae on day 6 pc, showing epithelial tissue with large numbers of secretory cells, cf. Fig. 5. Semithin section, Richardson stain. x 371

However, we can also see that the cytoplasm has stained very strongly, and that the oviductal constriction contains a considerably reduced number of secretion granules (Fig. 21). Staining capacity with alcian blue changes accordingly – both reactions (at pH 0.5 and pH 2.5) are obviously weaker in the preimplantation phase (Figs. 9b, 10b). Under the scanning electron microscope one finds that the buttonlike protuberances on the secretory cells have almost disappeared. Their surfaces appear smooth now, even though cell tops and protrusions are still visible; some of them contain even, fine-grained, dense granules, others are packed more loosely and infiltrated with secretion granules. This agrees with our light-microscopic observation of protrusions that seem to have separated off from the cells. In many cases we actually do find cell sections in the oviductal space which are completely surrounded by membrane (Fig. 22).

We have also observed club-shaped protrusions that are connected to the cell only by a thin strand of cytoplasm. The apices of secretory cells in the ampulla are characterized by large numbers of widely spaced cisterns of rER and by the presence of large secretion granules which are light in color and markedly different from those seen at estrus (Fig. 23).

In the oviductal constriction, hormonal effects on the secretory cells are much more evident. The secretory cells, which are long, sometimes slender or constricted

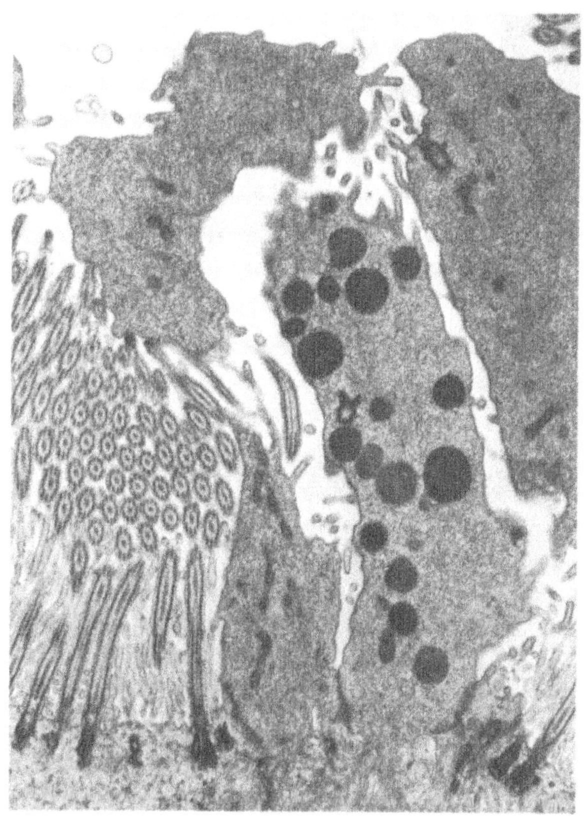

Fig 22. Pars ampullaris tubae on day 6 pc, showing secretory cell protuberances that have divided off from the cell, and two that are still connected with the cell body. On left and right, sections of ciliated cells. Electron micrograph. x 11,020

and sometimes broad, are packed in dense groups, and their polarity is still in evidence. As early as the 2nd day post coitum, however, a restructuring of cytoplasmic organization begins (Fig. 24). The entire supranuclear space takes on the appearance of a network or beehive. The cytoplasm is reduced to narrow strips and irregular peninsulas. In between we find secretory material of highly varying density. Of note is the fact that the membranes of numbers of secretory vacuoles have dissolved, allowing them to merge into broad lakes of secretion in which isolated membrane fragments are observed. Within these light-colored masses of secretion lie finite islands of secretory material that are more densely structured. Also apparent is the irregular, often fissured surface. In many cases the secretion granules approach the plasmalemma at the cell apex, losing their membrane in the process. In other cases the plasmalemma is ruptured and masses of secretion enter the oviductal space. In the remaining areas of cytoplasm we find small mitochondria, traces of granular ER, free ribosomes, and isolated small bubbles (Fig. 25). The cell nuclei are located in the lower third of the cell. They are elongated and often show several deep invaginations (Fig. 26). The chromatin is fine-grained and distributed fairly evenly throughout the cellular space with no apparent regional concentration. The perinuclear cytoplasm, however, does not give such a unified picture. Some of the cells have a very dense structure due to the presence of large numbers of very fine granules; others an extended rER of the most varying shape (Fig. 27). Particularly impressive are the hugely extended perinuclear cisterns, which

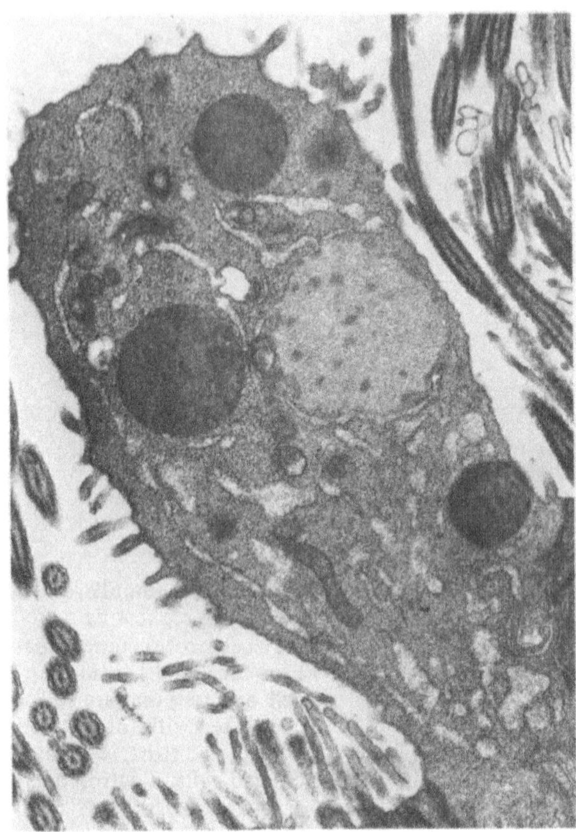

Fig. 23. Pars ampullaris tubae on day 6 pc, showing pear-shaped secretory cell protrusions with secretion granules of various density, mitochondria, and widely spaced rER cisterns. Electron micrograph. × 15,210

contain fine grains or flakes of low density material. We often observe large, blister-like spaces surrounded by membrane in the perinuclear cytoplasm; these may have resulted from a degeneration process. In the supranuclear area we find thick stacks of golgi membranes which are fully developed by day 6 post coitum.

Our morphological investigation of oviductal epithelium during the preimplantation phase revealed no structural changes in ciliary cells, however.

Concluding Remarks

A properly timed preimplantation phase is crucial to the normal development of mammalian embryos; optimal conditions for their nutriment, migration, and development must obtain in both oviduct and uterus. Under the influence of ovarian hormones the epithelium of oviduct and uterus produces a secretion that is absolutely necessary to the developing blastocyst and its implantation. This secretion, among electrolytes, amino acids, and metabolites for carbon dioxide metabolism, contains key proteins and glycoproteins (Beier 1966, 1967, 1968a, b, 1970, 1973, 1974, 1976, 1977; Kirchner 1969; Koester 1970; Petzoldt 1970; Schwick 1965; Stegner 1969; Petry et al. 1970; Kühnel et al. 1971). Since the mammalian blastocyst does not possess a large enough stock of nutrients of its own, it has to depend on the reservoir of oviductal and uterine secretion to fill its energy requirements up to the time of implantation.

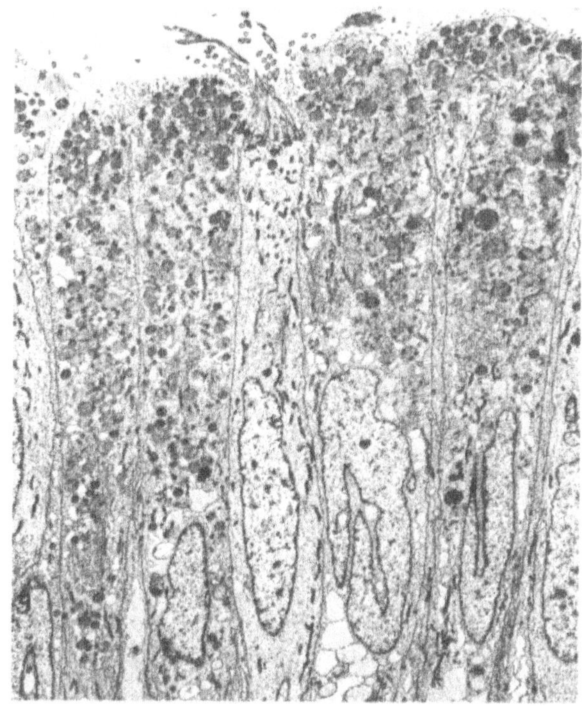

Fig. 24. Pars isthmica tubae on day 6 pc, showing highly prismatic, slender secretory cells with basally located nuclei and heteromorphic secretion granules. Note the ciliated cell in the center. Electron micrograph. × 3,175

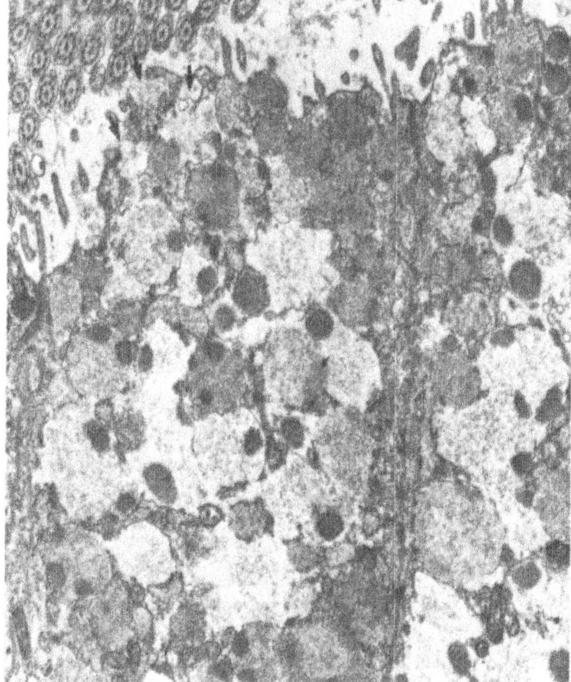

Fig. 25. Pars isthmica tubae on day 6 pc, showing apical sections of two secretory cells in which the secretion granules have lost their membranes and flow together. At (↗) note extrusion process. Electron micrograph. × 11,290

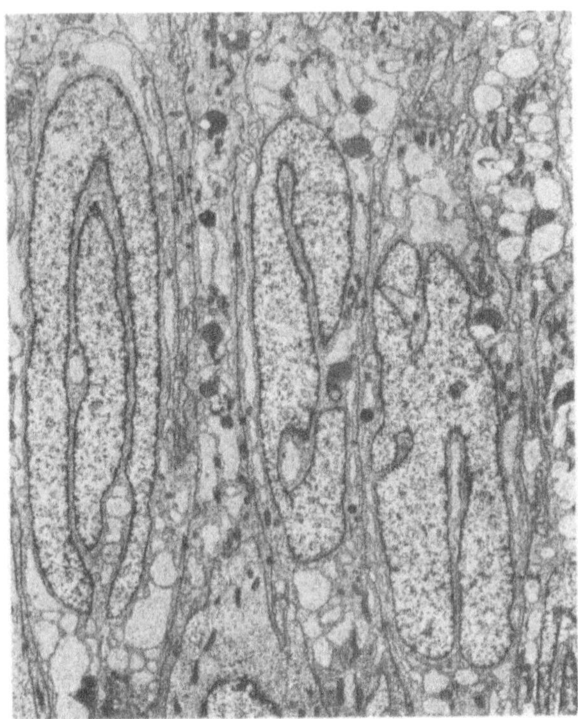

Fig. 26. Pars isthmica tubae on day 6 pc, showing nuclear region of secretory cells. Note the deep nuclear invagination. Electron micrograph. × 5,292

During its migration the rabbit blastocyst is surrounded by a second layer known as the mucoprotein layer (Denker 1970 a, b). Thus the process of migration is more than a simple passage through the oviduct to the site of implantation; the timing of this process is crucial if the blastocyst is to develop normally. This is why there have been so many, and different, suggestions as to the possible secretory transformation of oviductal epithelium. The present investigation, however, has shown that there is hardly any cytologic transformation of the endosalpinx epithelium at all during the preimplantation phase. The adaptation of the uterine system to blastocyst development and preparation for implantation is apparently more profound in nature than that of the oviductal epithelium to blastocyst migration (Beier and Kühnel 1976).

In the rabbit, only the secretion of mucoproteins seems of significance. The protein content of oviductal secretion does not change even with exogenous estradiol (Dies 1973). However, our studies indicate that, at least in the oviductal constriction, the secretory cells do indeed undergo changes that we can detect morphologically. Though in the ampulla, too, we have seen signs of directed secretion, these were limited to the release of secretion granules and the detachment of cell apices. In the secretory cells of the isthmus, by contrast, we have seen clear signs of cytoplasmic transformation. This transformation and secretion appears to be estrogen-dependent. Recent studies and analyses have shown that oviductal tissue contains large amounts of estrogen receptors, however hardly any for progesterone. The endometrium, on the other hand, has been shown to contain both estrogen and gestagen receptors (Beier and Kühnel 1976).

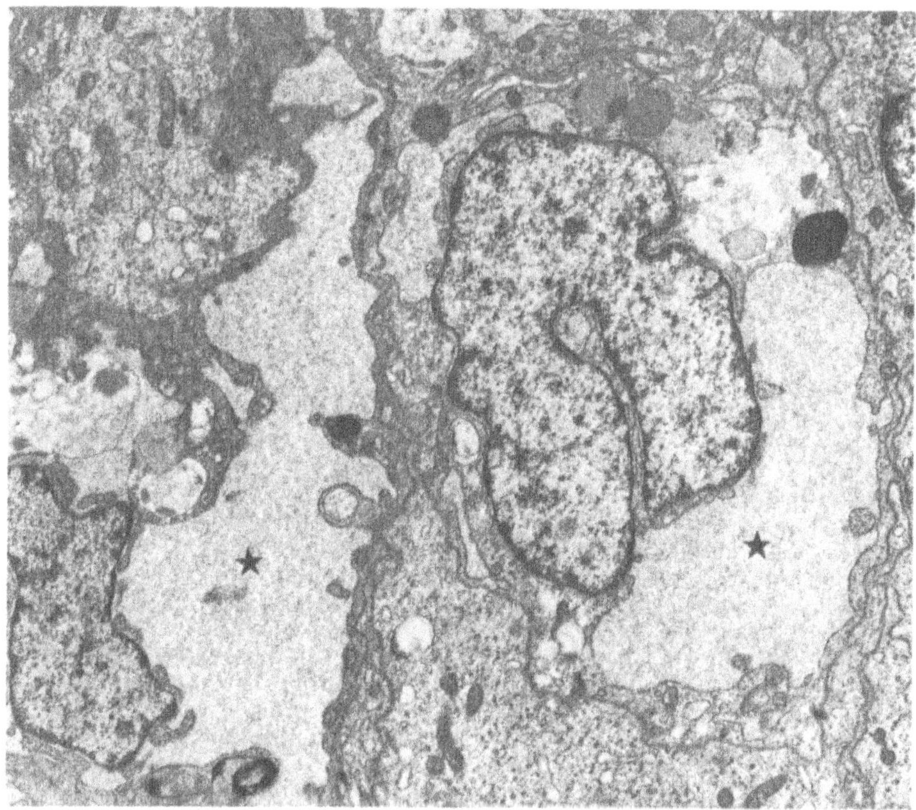

Fig. 27. Pars isthmica tubae on day 6 pc, showing basal sections of two secretory cells with greatly expanded cisterns of rER (↓). Electron micrograph. × 11,441

The studies that have been performed on the oviduct over the past few decades with light and electron microscopes have provided us with a tremendous amount of information about the cyclic transformation of oviductal epithelium and about how secretion is influenced by exogenous hormones. We have had statements about the disappearance of basophilia, i.e., the reduction of ergastoplasma in secreting cells, followed by development of the golgi apparatus and its preparation of secretion. This in turn is followed by extrusion, and then, cycle-dependent, a restituion of the ergastoplasma. Various opinions have also been expressed about the extent to which ciliary cells undergo cycle-dependent transformation. All of these conflicting findings should remind us that when we discuss hormone-controlled processes we must keep the total cell in mind, and that we cannot expect, by studying certain of its structural elements in isolation, to form a satisfactory picture of these life processes. Our observations should also serve to remind us that reliable results can be obtained only with methods that have been carefully checked and monitored in repeated experiments. In all of our extensive material, for example, we were not able to detect one instance of the "peg cells" within the oviductal epithelium, using careful and routinely perfusion fixing. However, they are often seen in immersion-fixed tissue. Whether or not the electron microscopy of the future will be able to show us the morphological equivalents of hormone-controlled processes, remains an open question.

Acknowledgements

These investigations were supported by Deutsche Forschungsgemeinschaft grant No. Ku 210/8.

I would like to thank Frau Ingeborg Ackermann, Frau Gabriele Bock, and Frau Gudrun Scheele for their conscientious help in preparing this paper.

References

Bargmann W (1977) Histologie und Mikroskopische Anatomie des Menschen. Thieme, Stuttgart

Beck LR, Boots LR (1974) The comparative anatomy, histology and morphology of the mammalian oviduct. In: Johnson AD, Foley CW (eds) The oviduct and its functions.. Academic Press, New York London, pp 1–51

Beier HM (1966) Das Proteinmilieu in Serum, Uterus und Blastocysten des Kaninchens vor der Nidation. In: Biochemie der Morphogenese, Chairman: W Beermann, Deutsche Forschungsgemeinschaft, Konstanz

Beier HM (1967) Veränderungen am Proteinmuster des Uterus bei dessen Ernährungsfunktion für die Blastocyste des Kaninchens. Verh Dtsch Zool Ges 31:139–148

Beier HM (1968a) Biochemisch-entwicklungsphysiologische Untersuchungen am Proteinmilieu für die Blastocystenentwicklung des Kaninchens (Oryctolagus cuniculus). Zool Jahrb Anat 85:72–190

Beier HM (1968b) Uteroglobin: A hormone-sensitive endometrial protein involved in blastocyst development. Biochem Biophys Acta 160:289–291

Beier HM (1970) Hormonal stimulation of protease inhibitor activity in endometrial secretion during early pregnancy. Acta Endocrinol (Kbh) 63:141–149

Beier HM (1973) Die hormonelle Steuerung der Uterussekretion und frühen Embryonalentwicklung des Kaninchens. Habilitationsschrift, Kiel

Beier HM (1974) Oviducal and uterine fluids. J Reprod Fertil 37:221–237

Beier HM (1976) Uteroglobin and related biochemical changes in the reproductive tract during early pregnancy in the rabbit. J Reprod Fertil [Suppl] 25:53–69

Beier HM (1977) Endocrinology and biochemistry of uteroglobin and other proteins involved in blastocyst development. Biochem Soc Trans 5:450–452

Beier HM, Kühnel W (1973) Pseudopregnancy in the rabbit after stimulation by human chorionic gonadotropin. Horm Res 4:1–27

Beier HM, Kühnel W (1976) Untersuchungen zur funktionellen Morphologie des Epithels der Endosalpinx und des Endometriums. Verh Anat Ges 70:831–838

Brenner RM (1969) The biology of oviductal cilia. In: Hafez ESE, Blandau RJ (eds) The mammalian oviduct. University of Chicago, Chicago London, pp 203–230

Brower LK, Anderson E (1969) Cytological events associated with the secretory process in the rabbit oviduct. Biol Reprod 1:130–148

Busch LC, Mootz U, Kühnel W (1977) Zur Oberflächenbeschaffenheit der Schleimhaut von Tube und Uterus des Kaninchens im Oestrus. Verh Anat Ges 71:525–529

Challis RRG, Davies IJ, Ryan KJ (1973) The concentrations of progesterone, esterone and estradiol–17β in the plasma of pregnant rabbits. Endocrinology 93:971–976

Denker HW (1970a) Topochemie hochmolekularer Kohlenhydratsubstanzen in Frühentwicklung und Implantation des Kaninchens. I. Allgemeine Lokalisierung und Charakterisierung hochmolekularer Kohlenhydratsubstanzen in frühen Embryonalstadien. Zool Jahrb Physiol 75:141–245

Denker HW (1970b) Topochemie hochmolekularer Kohlenhydratsubstanzen in Frühentwicklung und Implantation des Kaninchens. II. Beiträge zu entwicklungsphysiologischen Fragestellungen. Zool Jahrb Physiol 75:246–308

Dies R (1973) Hormonelle Beeinflussung des Sekretproteinmusters und der Keimentwicklung im Genitaltrakt von Oryctolagus cuniculus. Zool Jahrb Physiol 77: 360–441

Fawcett DW, Porter KR (1954) A study of the fine structure of ciliated epithelia. J Morphol 94:221–281

Fuchs AR, Beling C, Park K (1974) Corpus luteum recognition of blastocysts prior to implantation. Endocrinology 95:167

Greenwald GS (1969) Endocrinology of oviductal secretions. In: Hafez ESE, Blandau RJ (eds) The mammalian oviduct. University of Chicago, Chicago London, pp 183–202

Hafez ESE, Kanagawa H (1975) The female reproductive tract: Comperative aspects. In: Hafez ESE (ed) Scanning electron microscopy. Atlas of mammalian reproduction. Thieme, Stuttgart, pp 128–141

Hasan SH, Neumann F, Friedrich E, Elger W (1971) Levels of progesterone during pregnancy in rabbits as measured by radioimmunoassay. Acta Endocrinol (Kbh) [Suppl] 155:107

Hilliard J, Eaton LW (1971) Estradiol–17β, progesterone and 20 α-hydroxypregn-4-en-3-one in rabbit ovarian venous plasma. II. From mating through implantation. Endocrinology 89:522–527

Horstmann E, Stegner HE (1966) Tube, Vagina und äußere weibliche Genitalorgane. In: Möllendorf W von, Bargmann W (eds) Handbuch der Mikroskopischen Anatomie des Menschen Bd 7/Teil 4. Springer, Berlin Heidelberg New York

Kirchner C (1969) Untersuchungen an uterusspezifischen Glycoproteinen während der frühen Gravidität des Kaninchens (Oryctolagus cuniculus). Roux' Arch. Entw.-Mech. 164:97–133

Koester H (1970) Ovum transport. 21. Coll.Ges.Biol.Chem. Mosbach 1970. Springer, Berlin Heidelberg New York

Kühnel W, Busch LC (1979) Surface morphology of the rabbit uterus and oviduct during estrus. Anat Embryol (Berl) 156:189–195

Kühnel W, Beier HM, Petry G (1971) Untersuchungen zur hormonellen Regulation der Praeimplantationsphase der Gravidität. II. Histologische, topochemische und biochemische Analysen am hormonbehandelten Kaninchenuterus. Cytobiologie 4:9–40

Nilsson O, Reinius S (1969) Light and electron microscopic structure of the oviduct. In: Hafez ESE, Blandau RJ (ed) The mammalian oviduct. University of Chicago, Chicago London, pp 57–83

Pauerstein CJ, Eddy CA (1979) The role of the oviduct in reproduction; our knowledge and our ignorance. J Reprod Fertil 55:223–229

Petry G, Kühnel W, Beier HM (1970) Untersuchungen zur hormonellen Regulation der Praeimplantationsphase der Gravidität. I. Histologische, topochemische und biochemische Analysen am normalen Kaninchenuterus. Cytobiologie 2:1–32

Petzoldt U (1970) Bestimmung anorganischer Ionen in Uterus und Blastocyste des Kaninchens während der frühen Trächtigkeit. Zool Anz 33:128–133

Schwick HG (1965) Chemisch-entwicklungsphysiologische Beziehungen von Uterus zu Blastocyste des Kaninchens (Oryct. cuniculus). Roux' Arch. Entw.-Mech. 155: 283–343

Stegner HE (1969) Die funktionellen Wechselwirkungen zwischen Tube und befruchtetem Ei. Arch Gynaekol 207:136–154

Woodruff JD, Pauerstein CJ (1969) The fallopian tube. Williams Wilkins, Baltimore, MD

Morphological Findings in the Human Ovary Under Physiologic Conditions and After Contraceptive Use

W. Mestwerdt and D. Kranzfelder

Abstract

Systematic studies of the human ovary with the light and electron microscope have been used to define the granulosa cells and the theca cells in processes of follicle-ripening and atresia. We examined ovarian biopsies from 24 patients between 21 and 38 years of age, who were operated on for various reasons.
The follicles revealed the following structural qualities:
1. The stroma cells differentiated into steroid cells during the transition from the secondary follicle to the tertiary follicle containing liquor.
2. The granulosa cells, structurally differentiated as actively protein synthesizing cells, in the periovulation phase changed into steroid active granulosa-luteal cells of the corpus luteum.

In addition, measurement of the capillary density beneath the avascular granulosa layer served as a further structural parameter for determining processes of follicle ripening and atresia.

The results of our studies made of structurally normal ovaries under physiologic conditions are compared with the morphological findings of biopsies taken from women using oral contraceptives for up to 8 years, and these results are compared with those given in the literature.

There are three basic reasons for our interest in conducting functionally oriented, morphological studies of the human ovary. First, they might give us further insight into the anatomic and physiologic relationships between the separate tissue and cell types in the ovary during different times of a woman's life. Second, they should help us to define more precisely how these tissue and cell formations depend on intra- and extraovarian regulation mechanisms. Third, we wished to see whether and to what extent we could detect morphological changes in the ovary under the influence of exogenous substances (e.g., oral contraceptives, cytostatica).

Histologic changes in the ovary following oral contraceptive use have been described in detail by Ludwig and co-workers (1965, 1968). What electron microscopic studies have concentrated on to date has been the effect of low doses of hormonal contraceptives on the corpus luteum (Marquez-Monter et al. 1968). To our knowledge, no report has yet been published on combined light and electron microscopy of granulosa and theca cells in different follicle stages, under physiologic conditions, and following combined hormonal medication.

It is difficult to detect morphological changes in the ovary resulting from exogenous hormones, since under physiologic conditions there is already an intense interaction between follicle-ripening and -atresia. Processes of this kind cannot be described precisely in their initial stages, either in morphological or functional terms. That is why any structural findings in the ovaries after short or long-term hormonal medication should be interpreted with care.

Material and Methods

We obtained ovarian tissue from 17 patients who had regular menstrual cycles of between 28 and 30 days, and who had not taken hormonal medication. Our second sample comprised biopsies from the ovaries of seven patients who had been on combined medication anywhere from 6 months to 8 years (Tables 1 and 2). Tissue samples were invariably taken during the first half of menses, as a rule between the 10th and 15th day. We were particularly interested in the granulosa and theca cells of follicles in different stages of development. In order not to have to define ripening, and periovulating follicles solely by morphological criteria, blood samples were taken at 4 to 6 intervals over a period of at least 24 h before the operation, to allow us to determine levels of luteinization hormone. These levels were then correlated with morphological substrate.

Indication for operation, in most cases, was either a uterus myomatousus or a retroflexed uterus; more rarely a carcinoma in situ or invasive carcinoma of the cervix uteri.

A number of the biopsies were embedded in paraffin. The sections were dyed with HE, PAS, and Gomori. For electron microscopy the tissue was sawed into blocks with an edge length of 1 mm. They were then fixed in cold 4% glutaraldehyde, washed out in 0.1 M phosphate buffer, later fixed in 1% buffered osmium acid, and embedded in epon. Ultrathin sections were prepared using a Reichert OMU 3. After contrasting the sections with uranylacetate and lead citrate, we looked at them under an electron microscope (Zeiss EM 9 S). Precise orientation for ultramicrotomy was done with 1 μm thick, toloidin blue-dyed sections. We also counted mitoses in granulosa and theca cells per

Table 1. Patients on combined birth control medication

No.	Age	Medication	Length of therapy
1	38	Lyndiol	6 months
2	29	Anacyclin	1 year
3	21	Eugynon	1 year
4	29	Noracyclin	2.5 years
5	42	Anovlar	5 years
6	35	Eugynon	6 years
7	30	Eugynon	8 years

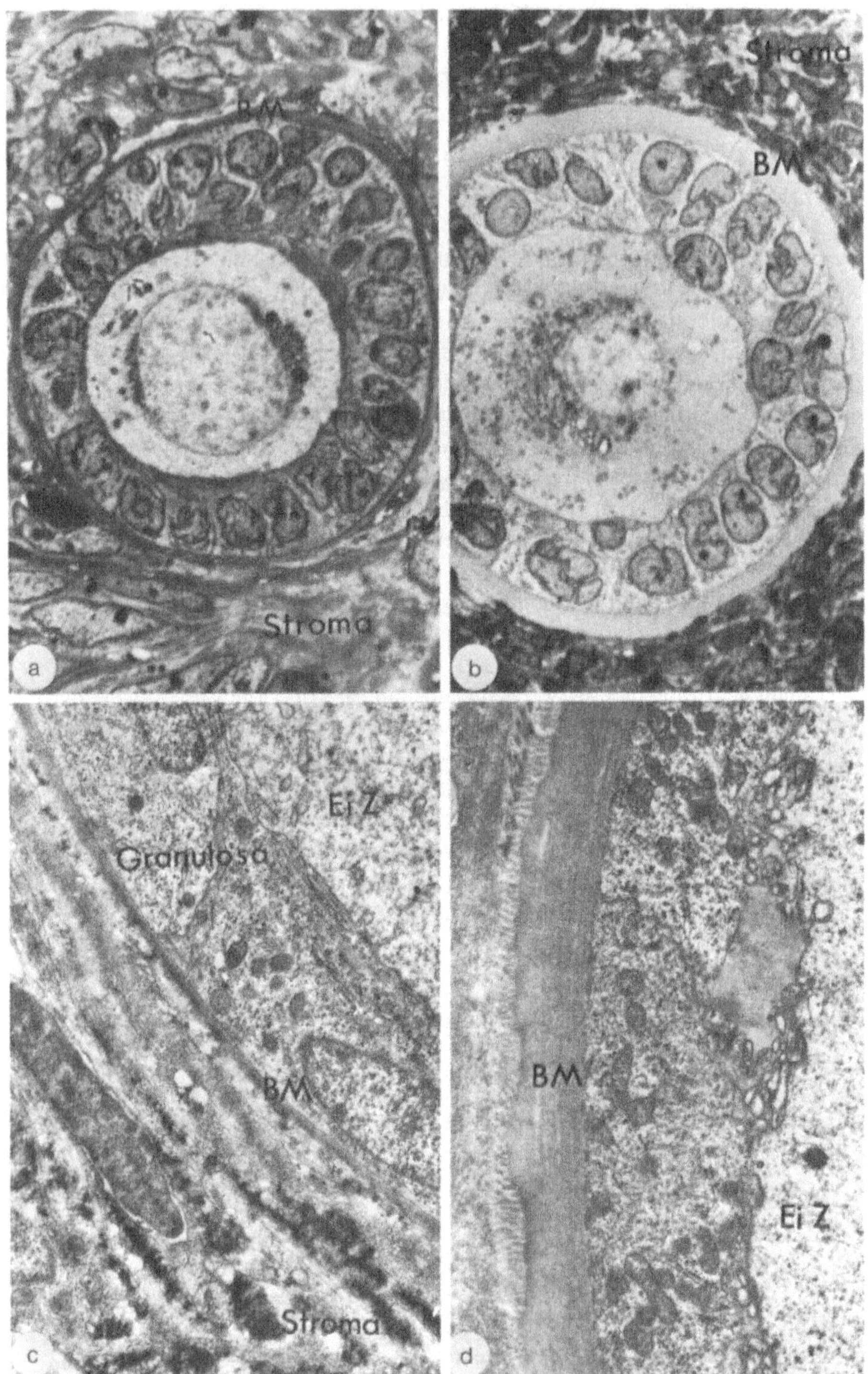

Fig. 1 a-d.

200 cells, as well as determining the vessel density on membrana granulosa by calculating the number of capillaries and vessels on a defined length of granulosa.

Histologic Findings

The main findings obtained with a light microscope on the follicles of both groups of patients are listed in Table 2 (see also Figs. 1 to 5). The follicles were classified on the basis of Stieve's (1943) and Watzka's (1957) findings.

We were able to follow the development of the follicles to the tertiary stage even in those patients on medication. This was true of all patients, even those who had been taking hormonal contraceptives for as long as 8 years.

With respect to the morphological parameters mentioned, we obtained similar structural findings for the primary follicles of both groups. However, there was a remarkable proliferation of the basal membrane around the granulosa of small follicles in the women on contraceptives. Histologically, this showed up very clearly as homogeneous material with intensively dyed eosinophils. In this case the thickness of the basal membrane was 6 μm (Fig. 1b). By comparison, the basal membranes around the granulosa of primary follicles in patients who had not used contraceptives were stretched very thinly (Fig. 1a).

Secondary follicles were observed only in the nonmedicated group; they were not present in the other group. Most of the tertiary follicles of the women who had used no oral contraceptives had diameters of up to 10 mm. Their granulosa contained three to five layers; the cells were mostly cuboidal and usually had a large nucleus with several nucleoli (Fig. 2a).

In the group on oral contraceptives we noted an extreme variation in the diameter of tertiary foliclcs. In one case we saw a follicle with a diameter of 18 mm, paired, however, with a vessel-free granulosa of only six layers; individual cells had already been released into the follicular chamber. Also worthy of note was a higher degree of vacuolization in the cytoplasm of these granulosa cells, and also in theca cells (Fig. 3a). Theca consisted of epitheloid cells and was criss-crossed by a dense network of vessels (Fig. 3a).

In the group of women who had a history of contraceptive use we also found one example of a ripening follicle with more than 12 cell layers in the granulosa region (Fig. 4a). Here, too, we noted large numbers of theca cells distributed in an epitheloid manner in the theca interna.

Graafian follicles could be identified by the strong proliferation of the associated granulosa, which generally had more than 16 cell layers (Fig. 5a). Event at this follicle stage the granulosa is still free of vessels. The theca interna is heavily vascularized and consists of large, epitheloid theca cells (Fig. 5a). This follicle group have been seen only in the group without contraceptive medication.

Fig. 1 a-d. Primary follicle. **a** Center-sectioned ovum surrounded by cuboid granulosa cells. Follicle separated from stroma by narrow basal membrane *(BM)*. × 1,102. **b** Basal membrane *(BM)* between granulosa and stroma thickened. From a patient who had taken Eugynon for 8 years. × 1,102. **c** Electron micrograph of stroma border zone, showing granulosa with 600 Å thick basal membrane *(BM)* × 4,014. **d** Electron micrograph showing laminations of basal membrane *(BM)* at the border between stroma and granulosa. From a patient who had taken Anovlar for 5 years. × 8,594

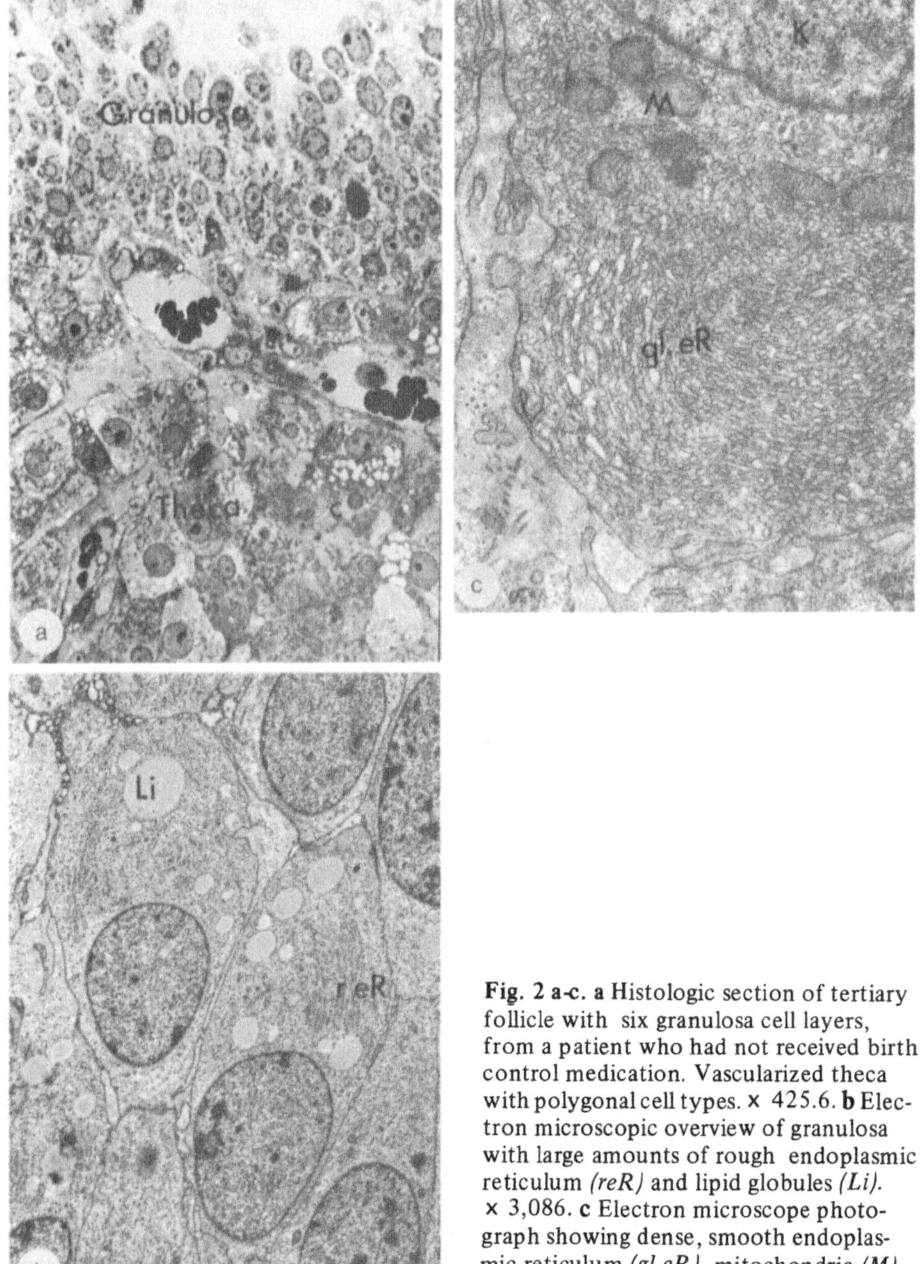

Fig. 2 a-c. a Histologic section of tertiary follicle with six granulosa cell layers, from a patient who had not received birth control medication. Vascularized theca with polygonal cell types. × 425.6. **b** Electron microscopic overview of granulosa with large amounts of rough endoplasmic reticulum *(reR)* and lipid globules *(Li)*. × 3,086. **c** Electron microscope photograph showing dense, smooth endoplasmic reticulum *(gl eR)*, mitochondria *(M)* and nucleus *(K)* in the theca cell. × 13,832

As can be seen from Table 2, particularly the follicles maturing in granulosa and theca showed a high incidence of mitoses. Also characteristic of follicle development was a changing vessel density in membrana granulosa, which varied from 1.6 mm at the primary to 54 μm at the graafian stage.

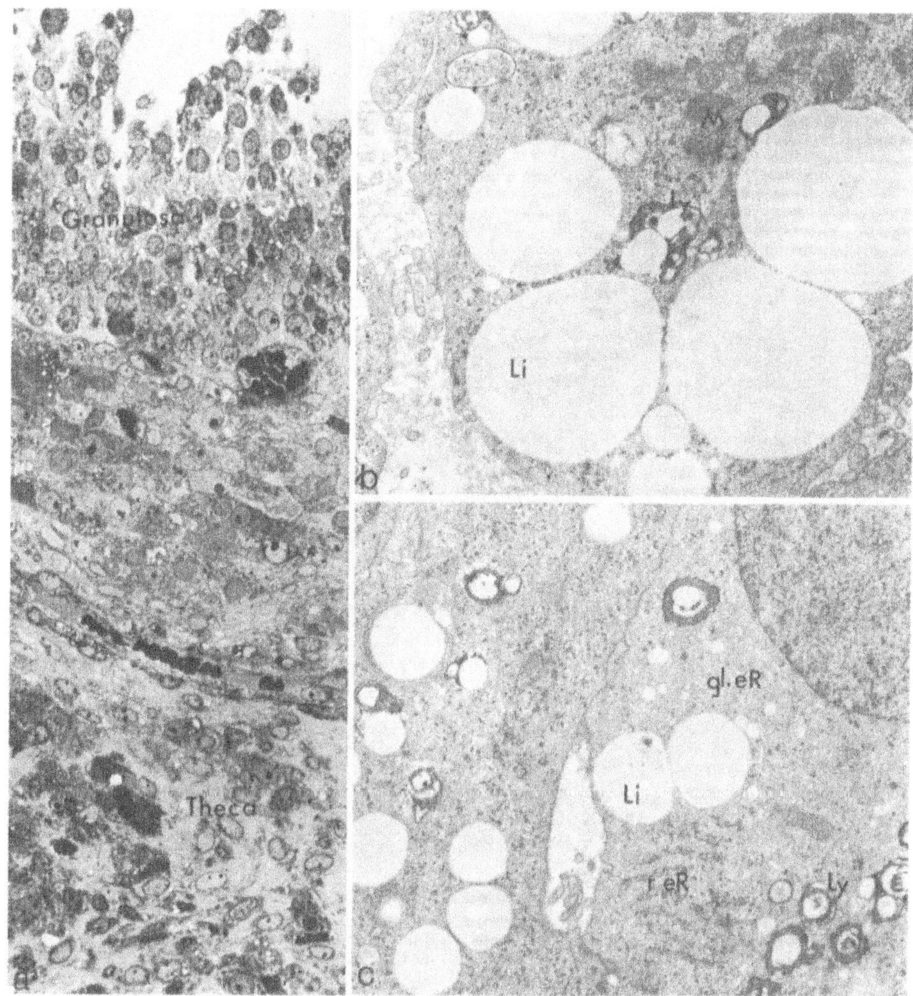

Fig. 3 a-c. a Histologic section of a tertiary follicle with eight granulosa cell layers, from a patient who had taken Eugynon for 1 year. Note the concentration of lipids in the cytoplasm, and vascularized *theca* with polygonal cell types. x 331.5. **b** Electron micrograph of a granulosa cell, showing large lipid globules and lysosome-like granula *Li* = lipid, *Ly* = lysosomes, *M* = mitochondria). x 6,465. **c** Electron micrograph of a theca cell showing lipid globules (Li) and lysosome-like granula *(Ly)*, *(eR* = endoplasmic reticulum in both rough *(r)* and smooth *(gl)* forms). x 6,465

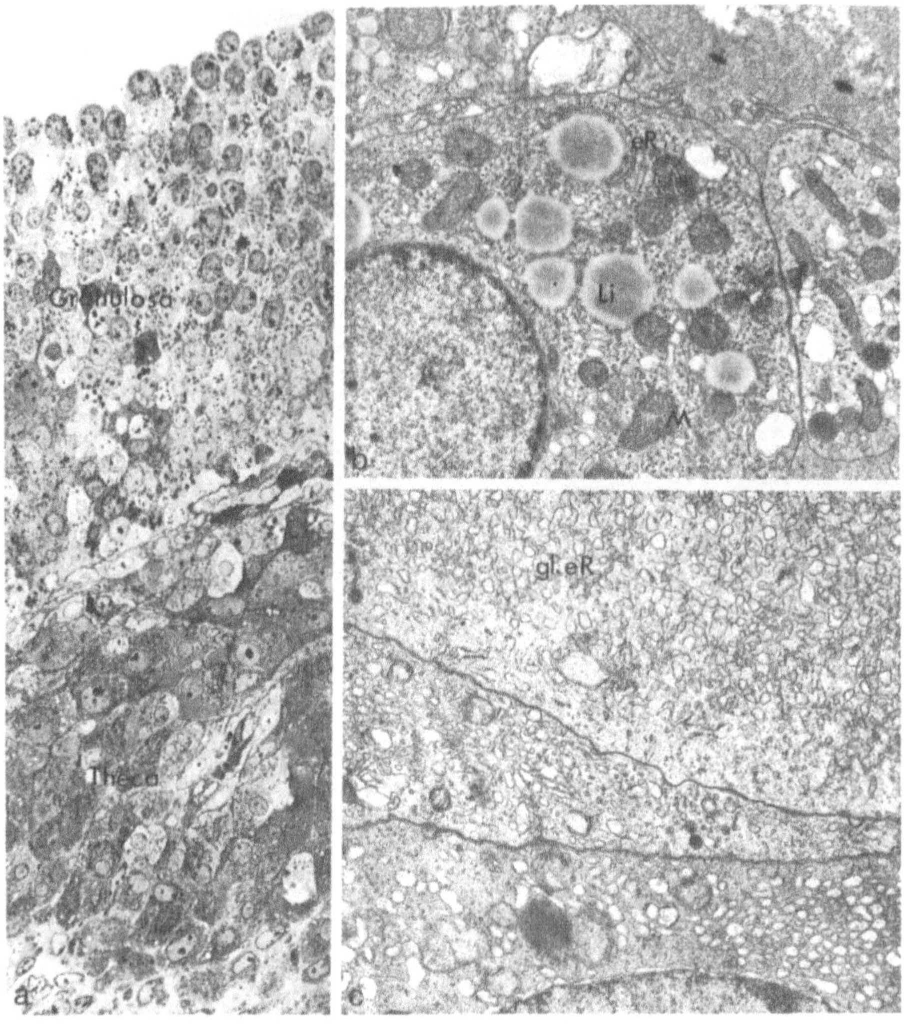

Fig. 4 a-c. a Detail of a ripening follicle with 12 layers of granulosa cells, from a patient who had taken Eugynon for one year then discontinued 6 weeks before operation. Note the vascularized *theca* with large polygonal cell types. x 348.3 **b** Electron micrograph of a granulosa cell. Note the endoplasmic reticulum in concentric laminations around the lipid globules *(Li)*, (*M* = mitochondria). x 11,320. **c** Electron micrograph of a theca cell, showing how the large area of its cytoplasm is filled with smooth endoplasmic reticulum *(gl. eR)*. It is lacking in other recognizable cytoplasmic organelles. x 14,990

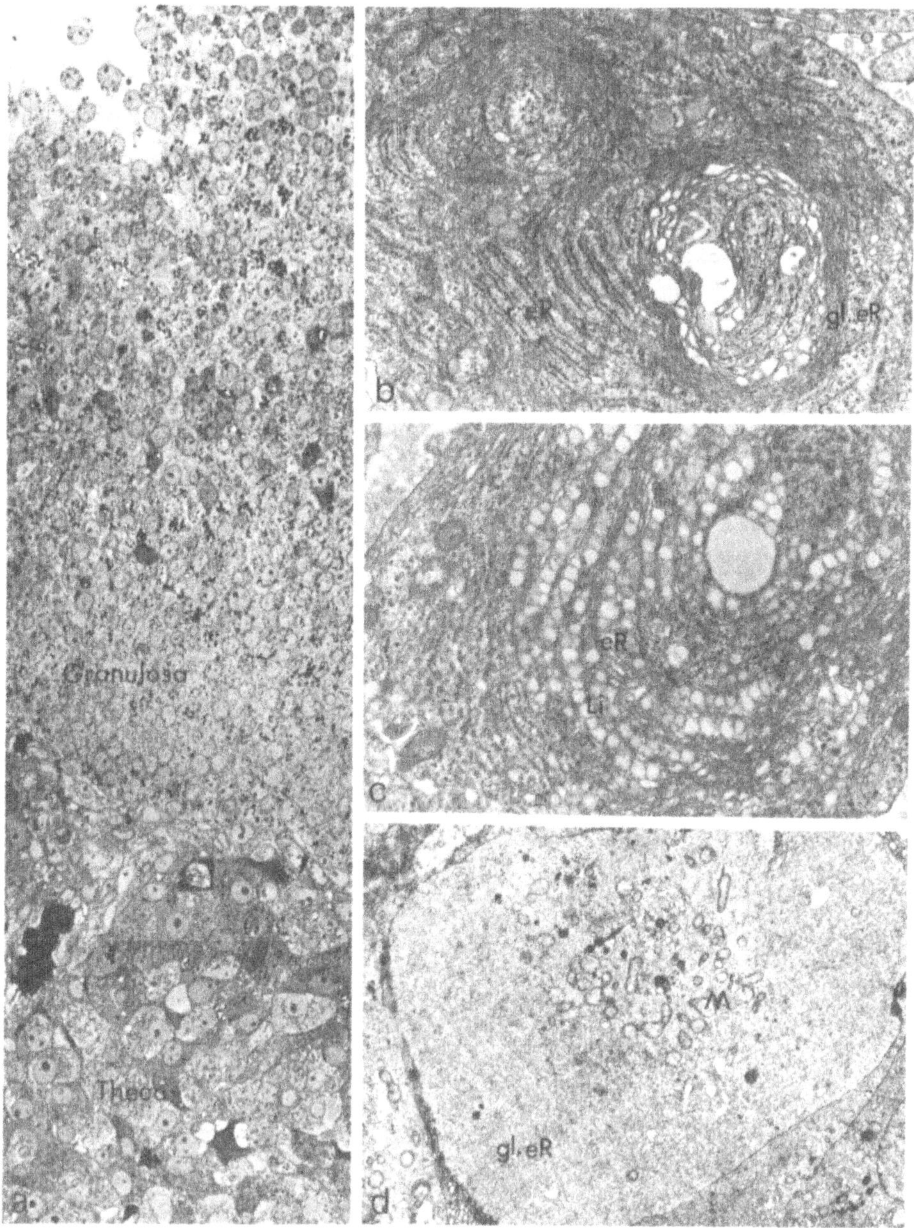

Fig. 5 a-d. **a** Histologic detail of a mature graafian follicle showing a granulosa with over 16 cell layers. From a patient who had not received birth control medication. Vascularized theca with polygonal cell types. x 343.8. **b** Gradual transition from rough *(r)* to smooth *(gl)* endoplasmic reticulum *(eR)*. x 14,797. **c** Lipid globules *(Li)* like strings of pearls between laminations of endoplasmic reticulum *(eR)*. x 15,411. **d** Theca cell with primarily smooth endoplasmic reticulum *(gl.eR)*, very little lipid and few other organelles (*M* = mitochondria). x 2,824

Table 2. Histologic characteristics of follicles at various developmental stages (modified after Stieve 1943 and Watzka 1957)

	Number of follicles	Size	Granulosa	Theca	Mitose-rate / 200 cells		Distance between 2 vessels at membranegranulosa
					Granulosa	Theca	
Group I: 17 women not using contraceptives							
Primary follicle	26	40–80 μm	one layer (flat-cuboid)	–	–	–	≈ 1.8 mm
Secondary follicle	3	100–800 μm	3–5 layers (cuboid)	–	0–1	–	560 μm
Tertiary follicle	5	900–10,000 μm	2–7 layers (cuboid)	interna et externa (polygonal cellform)	0–1	0–1	185 μm
Ripening follicle (taken at LH rise)	2	10–14 mm	8–18 layers (cuboid-polygonal)	interna et externa (polygonal cell form)	1–2	0–2	64 μm
Graafian follicle (taken at LH peak)	3	> 15 mm	>16 layers	interna et externa (polygonal cell form)	0–1	0–1	53 μm
Group II: 7 women using contraceptives							
Primary follicle	116	30–80 μm	one layer (flat-cuboid)	–	–	–	1.6 mm
Secondary follicle	–	–	–	–	–	–	–
Tertiary follicle (quiescent? atretic?)	6	200–18,000 μm	2–8 layers	interna et externa (polygonal cell form)	0–1	0–1	220 μ
Ripening follicle	1	14 mm	12 layers	interna et externa (polygonal cell form)	1–2	1–2	54 μ
Graafian follicle	–	–	–	–	–	–	–

Electron Microscopic Findings

The findings obtained with an electron microscope on primary follicles were comparable to the above for both patient groups, in terms of submicroscopic structure of ovum, granulosa, and the stroma surrounding the immature follicle (Figs. 1c and d). An interesting finding was a proliferation of the basal membrane around the primary follicles in patients who had a history of oral contraceptive use. Seen under the electron microscope, this membrane appears not as a homogeneous mass but as a layer of laminated basal membranes (Fig. 1d). Secundary follicles have not been observed in this group.

The granulosa of tertiary follicles of patients not on contraceptives had a large amount of rough endoplasmic reticulum in cytoplasm. We also detected fat globules of various sizes. The nuclei often contained several nucleoli (Fig. 2c). The cytoplasm of theca cells at this follicle stage elicit, under the electron microscope, a dense, membranous system of endoplasmic reticulum (Fig. 2d).

Interesting were the findings in granulosa and theca of tertiary follicles in women on birth control medication: In granulosa we found extremely large vacuoles, a sign of dissolved lipids. Further, we detected in granulosa cells a large number of what are often called lysosomelike granula (Fig. 3b). The theca cells were also rich in large globules of lipid vacuoles and lysosome-like granula (Fig. 3c).

The ripening follicles of both groups of patients showed similar structures, both under light and electron microscopes. Very conspicuous in the cytoplasm of granulosa cells were laminations of endoplasmic reticulum deposited around individual fat globules. Not infrequently in these granulosa cells we observed mitochondria of the vesicular type which is characteristic of steroid cells (Fig. 4b). When the theca cells have enlarged, one can recognize in their cytoplasm a very dense and membranous, smooth endoplasmic reticulum; however, other cytoplasmic organelles were not detectable.

We performed electron microscopy only on granulosa and theca of graafian follicles from patients not on birth control medication. In the cytoplasm of many of these cells we saw a gradual transition of rough endoplasmic reticulum to smooth. The lipid was increasingly finely distributed by progressive follicle stages, and sometimes lay like strings of pearls between the laminations of the smooth endoplasmic reticulum (Figs. 5b and c).

Figure 5d is an electron micrograph of a theca cell in a graafian follicle; practically the entire cytoplasmic space is filled with dense, membranous, smooth endoplasmic reticulum.

Evaluation of Findings

Like other investigators were able to observe, in ovarian tissue from women on combined hormonal birth control medication, the ripening of follicles up to the tertiary stage (Ludwig 1965, 1968; Weber 1969; Mall-Haefeli et al. 1969; Starup and Visfeldt 1974; Gondos 1976). The last named author has stated that the number of primary follicles was normal in ovarian tissue following medication, but that the number of tertiary follicles was reduced. In our own sample we were able to detect instances of such follicles even after 8 years of Eugynon therapy. The membrana granulosa in this case was markedly thin, having seven cell layers at the most. The cytoplasm showed frequent vacuolization.

In contrast to the findings of Ludwig (1968) and Mall-Haefeli et al. (1969) our samples also elicited a heavily vascularized theca with large, polygonal, epitheloid-layered theca cells. The large numbers of vacuoles in the cell cytoplasm were, however, impressive. Ludwig (1968) and Mall-Haefeli (1969) also report that in place of theca interna an uncharacteristic, granulated tissue develops around the granulosa of tertiary follicles in women using oral contraceptives. Corpora lutea, according to Erb and Ludwig (1965), Ludwig (1968), and Mall-Haefeli (1969) detectable in the ovaries of patients with histories of contraceptive use, were not observed in our sample.

Electron microscopy of granulosa and theca of various follicle stages in women on medication revealed some remarkable structural differences not detectable physiologically. At the primary follicle stage the findings agreed in terms of the structure of ovum , granulosa, and stroma surrounding the small follicle. However, the broad basal membrane around the granulosa of primary follicles, which looked like a homogeneous mass under the light microscope, appeared under the electron microscope as a wide band of laminated basal membranes – a finding that we have not seen reported in the literature in this form.

In the tertiary follicles of women on oral contraceptives we often detected large vacuoles and lysosome-like granula, in the cytoplasm of both granulosa and theca cells. These lysosome-like granula were present mostly in follicles with advanced atresia, and were an obligatory cytoplasmic organelle in the sample group that had no history of contraceptive use. Both light an electron microscopy showed that most of the tertiary follicles of patients on contraceptives behaved like atretic follicles, whereby we were not able to determine by purely morphological criteria the exact degree of atresia. In the samples investigated, the signs of atresia were not equally developed in all areas of the follicular tissue.

In none of our patients with a history of contraceptive use were we able to observe those remarkable structural transformations of granulosa cells from protein to steroid synthesis in tertiary and ripening follicles (cf. Mestwerdt et al. 1977 a, b). Only in one woman who had taken Eugynon for a year and then stopped 6 weeks before her operation, did our light an electron microscopic findings on granulosa and theca cytoplasm compare to those on ripening follicles taken under physiologic conditions. This example might be seen as evidence for the ability of follicles to rapidly regenerate after contraceptives are ceased. It has been corroborated by several other authors (Ludwig 1968; Mall-Haefeli et al. 1969; Weber 1969; Maqueo et al. 1972; Starup and Visfeldt 1974; Gondos 1976). Whether one can expect follicles to regenerate so quickly even after years of hormonal birth control therapy, is a question we could not answer on the basis of our sample.

In sum we may say, in agreement with most reports in the literature, that even after long-term use of oral contraceptives one can observe follicles ripening to the tertiary stage. Both light an electron microscopy revealed signs of atresia in various follicle stages in ovarian biopsies from women with a history of oral contraceptive use. Under the electron microscope this showed up particularly in a thickening of the basal membrane around the primary follicle. In tertiary follicles we often found, in addition to large lipid globules, lysosome-like granula in the cytoplasm of granulosa and theca cells. After cessation of short-term, if not long-term hormonal medication, our findings indicated that the structural and functional processes of the ovary went rapidly back to normal.

References

Erb H, Ludwig KS (1965) Strukturelle und funktionelle Veränderungen am menschlichen Ovar unter Einwirkung hormonaler Antikonzeptiva. Experientia 21: 159–162

Gondos B (1976) Histologic changes associated with oral contraceptive usage. Ann Clin Lab Sci 6: 291–299

Ludwig KS (1968) Zur funktionellen Morphologie des menschlichen Ovariums unter Einwirkung von Oestrogen-Gestagen-Kombinationspräparaten. Gynaekol Rundsch 6: 241–251

Mall-Haefeli M, Ludwig KS, Keller M, Weber W (1969) Beitrag zum Wirkungsmechanismus der oralen Ovulationshemmer beim Menschen. Geburtshilfe Frauenheilkd 171: 1–12

Maqueo M, Rice-Wray E, Calderon JJ, Goldzieher JW (1972) Ovarian morphology after prolonged use of steroid contraceptive agents. Contraception 5/3: 172–185

Marquez-Monter H, Gutierrez-Najar A, Aznar R, Ginervelazquez J, Rudel H, Martinez-Manautou J (1968) Ultrastructure of corpus luteum during low-dose contraceptive therapy Fert and Sterility 19: 363–371

Ludwig KS (1965) Über die morphologischen Veränderungen am menschlichen Ovar unter Einwirkung eines hormonalen Antikonzeptivum. Experientia 21: 726–728

Mestwerdt W, Müller O, Brandau H (1977a) Die differenzierte Struktur und Funktion der Granulosa und Theka in verschiedenen Follikelstadien menschlicher Ovarien. 1. Mitteilung: Der Primordial-, Primär-, Sekundär- und ruhende Tertiärfollikel. Arch Gynaekol 222: 45–71

Mestwerdt W, Müller O, Brandau H (1977b) Die differenzierte Struktur und Funktion tion der Granulosa und Theka in verschiedenen Follikelstadien menschlicher Ovarien. 2. Mitteilung: Der reifende, reife, sprungreife und frisch geplatzte Follikel. Arch Gynaekol 222: 115–136

Starup J, Visfeldt J (1974) Ovarian morphology and pituitary gonadotrophins in serum during and after long-term treatment with oral contraceptives. Acta Obstet Gynecol Scand 53: 161–167

Stieve H (1943) Über Follikelreifung, Gelbkörperbildung und den Zeitpunkt der Befruchtung beim Menschen. Z Mikrosk Anat Forsch 53: 467–582

Watzka M (1957) Weibliche Genitalorgane. Das Ovarium. In: Bargmann W (ed) Handbuch der mikroskopischen Anatomie des Menschen, vol VII/3. Springer, Berlin Göttingen Heidelberg, pp 1–178

Weber W (1969) Zur funktionellen Morphologie des menschlichen Ovariums nach Absetzen von Östrogen-Gestagen-Kombinationspräparaten. Geburtshilfe Frauenheilkd 29: 149–159

The Frequency of Spontaneous Abortion After Discontinuance of Oral Contraceptives

Gisela Dallenbach-Hellweg

We have observed over the past few years a sharp increase in the number of spontaneous abortions in which the embryo was abnormally developed, and we began searching our patients' medical histories for possible causes. Realizing that either primary damage to the germ cell or the inability of the endometrium to act as a nidation bed can lead to these symptoms, we focused our attention on the various drugs that have a noxious effect on ovary and endometrium. The logical step, of course, was to look at oral contraceptives, the consumption of which has risen very rapidly over the past 10 years.

Long-term use of oral contraceptives has been shown to lead to chronic hormonal imbalance, villous fibrosis in the ovary, and a degeneration and decrease in number of primary follicles (Ryan et al. 1964; Diddle et al. 1966; Plate 1967; Starup 1967; Zussman et al. 1967; Puyol-Amat et al. 1969; Dallenbach-Hellweg 1972). Frequently, polycystic changes that resemble an acquired Stein-Leventhal syndrome have been observed. Cytogenetic studies have shown what the degeneration of primary follicles can lead to: In women who had conceived within 6 months of discontinuing oral contraceptives and then suffered abortions, chromosome anomalies were found almost three times more frequently than in randomly selected cases. Topping the list by far were polyploidies, at 30% compared to 5% for the controls (Carr 1969, 1970; see also Mazo and Abrisqueta, 1979). Histologic and embryologic studies have shown that early embryonic abnormalities ("abortive ova") are found twice as frequently in women who had taken contraceptives than in those who had not (Poland 1970). Chromosome anomaly suggests a disturbance in ovum maturation, and particularly in meiosis, which under the influence of progesterone follows a very subtle time plan. The hormonal imbalance that persists even after oral contraceptives have been discontinued can be expected to disturb this coordinated process, so that if conception takes place abnormal chromosome constellations will probably occur (Knörr and Haas 1971). Other authors have found different results, but the drugs and constellations they looked at were also different; thus their findings do not necessarily refute the above statement, which has been corroborated elsewhere.

Drugs high in gestagens, when used over a long term, lead to atrophy of the endometrium, and drugs high in estrogens lead to hyperplasia of varying degrees. Experience has shown that both types of drug render the endometrium unfit for nidation. The time it takes for an atrophic endometrium to regenerate after the drug is discontinued will depend on the composition of that drug (Ferin 1964).

Our *own investigations* were based on a group of 200 women who had suffered abortions after discontinuing contraceptives, and a control group of equal size. We began by eliminating from the sample those cases for which either clinical history or material was insufficient. We considered only those cases for which type of drug, length of time taken, and interval between discontinuance and onset of pregnancy were known, and in which we had intact, i.e., nonnecrotic, placental tissue *and* endometrium for histologic evaluation. This reduced the number of cases to 113 precisely definable abortions following contraceptive use. Our control group of women who had had abortions but had not taken contraceptives, by the same token, finally numbered 100. The ages of the women in both groups corresponded almost one to one, with a tolerance of ± 3 months.

In terms of our patients' medical history, we began on the assumption that the appearance of pathologic changes would stand in direct relation to the interval between discontinuance of contraceptive and date of conception. Thus we divided our patients into three groups by length of interval: zero to 6 months, 7 to 12 months, and 12 months and over. We also classified the patients by type of substance taken (i.e., gestagen-high, estrogen-high, and neutral drugs) as well as by individual drug; we also compared the groups in terms of pathologic alterations in placenta and endometrium.

In histologic terms, we registered on the one hand deviations from normal placentation, in particular changes in the diameter of villi, in trophoblast epithelium, in villous stroma, and in fetal vascularization. On the other hand we recorded the development and functional state of the endometrium.

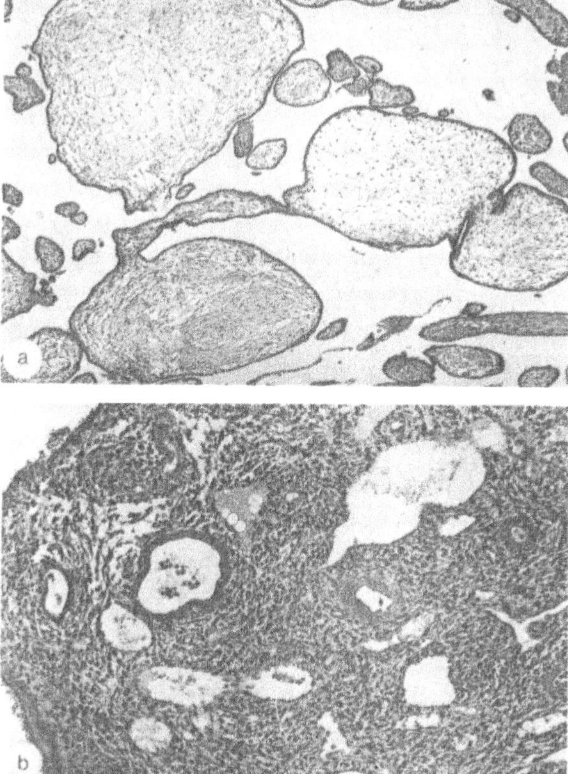

Fig. 1. Spontaneous abortion at fourth month, after discontinuing oral contraceptives which had been taken continuously for six years. **a** Malformed avascular hydropic villi, **b** the associated endometrium irregularly and deficiently developed

Table 1. Placental changes

Oral contraceptive group						
No. of Interval (months)	No. of Patients	Pathologic changes	Villous hydropathy	Villous fibrosis	Deficient fetal vascularization	Atrophy of trophoblast
0–6	82	38 (46%)	30 (37%)	12 (15%)	31 (38%)	20 (24%)
7–12	8	6	5	1	6	4
> 12	23	9 (40%)	6	4	8	5
Gestagen-high drugs	67	57%	45%	22%	48%	27%
Estrogen-high drugs	10	33%	33%	–	33%	–
Neutral drugs	34	30%	20%	9%	27%	21%
Totals	113	47%	37%	15%	40%	26%
Control group						
	100	21%	16%	3%	17%	14%

Table 2. Endometrial changes

Oral contraceptive group					
Interval (months)	No. of patients	Pathologic changes	Atrophy	Irregular proliferation	Abortive secretion
0–6	82	32 (39%)	21 (26%)	10 (12%)	4 (5%)
7–12	8	2	2	–	1
> 12	23	6 (26%)	3	3	2
Gestagen-high drugs	67	28 (42%)	16 (24%)	8 (12%)	5 (8%)
Estrogen-high drugs	10	1 (10%)	–	–	1
Neutral drugs	34	10 (30%)	7	5	1
Totals	113	35%	23%	12%	6%
Control group					
	100	3%	–	3%	–

Our *results* are detailed in Tables 1 to 4 and Fig 1. Disturbances in placentation in the contraceptive group very often took the form of increased villus diameter, hydropic swelling of villous stroma to the point of foci of degeneration and some fibrosis, insufficient or missing fetal vascular growth, and atrophy of trophoblast epithelium. Retarded endometrium development was characterized by serious glandular atrophy, stromal transformation (decidualization, spindle cells), and low overal thickness. We also noted local irregularities in the development of glands and stroma.

An evaluation of our findings leads us to stress the following points as being particularly important:

1. Abnormal development of placenta and endometrium, in terms both of frequency and degree, does not depend so much on the interval between discontinuance of contraceptive use and the date of conception, as on the composition of the drug. Even after intervals of 12 months and longer placental abnormalities were seen in a high percentage of cases – but only, however, in women who had been taking drugs high in gestagens. These findings are easily explained on the basis of the endometrial atrophy that often occurs after use of these drugs and that is so difficult to reverse, and that very probably prevents an earlier implantation. By contrast, one would assume that a short interval would lead to less serious, and more easily reversible, damage to ovary and endometrium. That these highly potent gestagens have a particularly damaging effect was also shown by Kühne et al. (1972), 17 of whose 18 patients on chlormadinone suffered embryonal abnormalities. The divergent results of other authors (Dhadial et al. 1970; Boue et al. 1975; Lauritzen 1975; Klinger et al. 1976) are probably explained by the various compositions of the contraceptives they tested.
2. In the group on contraceptives, pathologic changes in the placenta were found more often than underdeveloped endometrium. If one considers, however, that some of these placental abnormalities were probably secondary results of an underdeveloped nidation bed, the difference between these findings is obscured. All the same, the group of 67 women who had taken contraceptives high in gestagen contained 13 cases where placental damage was associated with normal endometrium, and only two cases in which though the endometrium was underdeveloped, the placenta was normal. These

Table 3. Pathologic findings: correlation between interval and drug combination

	Interval (months)	Gestagen-high	Estrogen-high	Neutral
Placenta	0 – 6	48%	3 of 8	42%
	7 – 12	6 of 7	–	–
	> 12	9 of 12	–	–
Endometrium	0 – 6	44%	1 of 8	38%
	7 – 12	2 of 7	–	–
	> 12	5 of 12	–	1 of 8

Table 4. On the frequency of pathologic changes in spontaneous abortion after discontinuing oral contraceptive agents

Pathologic changes	Oral contraceptive group	Control group
Placenta:		
Hydropic swelling	37%	16%
Deficient vascular growth	40%	17%
Atrophy of trophoblast	26%	14%
Villous fibrosis	15%	3%
Endometrium:		
Atrophy	23%	–
Irregular proliferation	12%	3%
Abortive secretion	6%	–

13 cases, all of them women who had taken drugs containing norgestrel, almost certainly involved damage to the embryo caused by the ovary.

In sum, we know from experience that both disturbances in endometrial function and primary embryonic damage can lead to spontaneous abortion. If both are present, embryonic damage may be secondary, i.e., a healthy blastocyst will degenerate after nidation in an underdeveloped endometrium. In other words, only if the endometrium is found to be normal, can we say with certainty that the developing blastocyst has been damaged by ovarian dysfunction.

In our sample, the disturbances in endometrial function that led to spontaneous abortion reached their peak within 6 months after contraceptives were discontinued. Placental changes, on the other hand, continued to be observed quite often even after 12 months. This would seem to justify the conclusion that the endometrium recovers from the effects of contraceptives more rapidly than does the ovary. Ovarian damage, which persists longer and hence must be taken more seriously, was found primarily with drugs high in gestagens, and particularly those containing the highly potent norgestrel. It would not seem far fetched to assume that this substance disrupts ovarian function specifically and lastingly.

Two practical conclusions may be drawn from our results. First, we ought to recommend to women who have been taking neutral or estrogen-high drugs to let at least 6 months go by from their last pill to conception; and secondly, drugs that are high in gestagens, and particularly norgestrel, ought only to be prescribed in exceptional cases to women who still wish to have children.

References

Boué J, Boué A, Zazar P (1975) The epidemiology of human spontaneous abortions with chromosomal anomalies, aging gametes. Blandau RJ (ed) Karger, Basel

Carr DH (1969) Genetic factors in pregnancy wastage. Med Clin North Am 53:5

Carr DH (1970) Chromosome studies in selected spontaneous abortions 1. Conception after oral contraceptives. Can Med Assoc J 103:343

Dallenbach-Hellweg G (1972) Therapieschäden in der Gynäkologie (Morphologische Beobachtungen über die Auswirkungen weiblicher Sexualhormone und von Stoffen mit ähnlicher Wirkung). Verh Dtsch Ges Pathol 56:252

Dhadial RK, Machin AM, Tait SM (1970) Chromosomal anomalies in spontaneously aborted human fetuses. Lancet II:20

Diddle AW, Watts GF, Gardner WH, Williamson PJ (1966) Oral contraceptive medication: a prolonged experience. Am J Obstet Gynecol 95:489

Ferin J (1964) Hypoestrogenic amenorrhoea and/or sterility induced by lynestrenol. Int J Fertil 9:29

Klinger HP, Glasser M, Kava HW (1976) Contraceptives and the conceptus. I. Chromosome abnormalities of the fetus and neonate related to maternal contraceptive history. Obstet Gynecol 48:40

Knörr K, Haas B (1971) Cytogenetische Untersuchungen bei Einnahme von Ovulationshemmern. In: Kewitz H (Hrsg) Nebenwirkungen contraceptiver Steroide. Westkreuz, Berlin, p 116

Kühne D, Seidl S, Göretzlehner G (1972) Contraceptive treatment with chlormadinone and its effect on the endometrium. A histological investigation. Endokrinologie 59:295

Lauritzen JG (1975) The significance of oral contraceptives in causing chromosome anomalies in spontaneous abortions. Acta Obstet Gynecol Scand 54:261

Del Mazo J, Abrisqueta JA (1979) Maternal origin of a trisomy 7 in a spontaneous abortion. Obstet Gynecol 53:185

McQuarrie HG, Scott CD, Ellsworth HS, Harris JW, Stone RA (1970) Cytogenetic studies on women using oral contraceptives and their progeny. J Obstet Gynecol 108:659

Plate WP (1967) Ovarian changes after long-term oral contraception. Acta Endocrinol (Kbh) 55:71

Poland BJ (1970) Conception control and embryonic development. Am J Obstet Gynecol 106:365

Pujol-Amat P, Urgell-Roca JM, Esterben-Altirriba J, Marquez-Ramirez M (1969) Studies of ovarian biopsies from women cyclically treated with the combination ethynodiol-diacetate and mestranol alone and together with human menopausal gonadotropin (Pergonal) + HCG. Gynaecologica 167:237

Ryan GM, Craig J, Reid DE (1964) Histology of the uterus and ovaries after long-term cyclic norethynodrel therapy. Am J Obstet Gynecol 90:715

Starup J (1967) The effect of gestagen and oestrogen treatment on the development of ovarian follicles: laboratory observations. Acta Obstet Gynecol Scand [Suppl 9] 46:15

Zussman WK, Forbes DA, Carpenter RJ Jr (1967) Ovarian morphology following cyclic norethindrone-mestranol therapy. Am J Obstet Gynecol 99:99

Combined Exogenous Hormones and Ovarian Cancer

F. D. Dallenbach

Because new facts indicate human ovarian carcinoma is somehow related to hormonal imbalances, I have elected to discuss these relationships, to present new information, hoping thereby to incite discussion and to promote exchange of information.

In America and in Great Britain ovarian carcinoma in women has become the most lethal primary genital tract neoplasm (Lingeman, 1974, Editorial-2, 1978). It causes more deaths than cervical or endometrial cancer combined. Autopsy records of a 100 years ago, however, indicate that ovarian cancer was extremely rare.

To try to find reasons for the increase in incidence of ovarian cancer and to learn what factors predispose to the disease, epidemiological surveys in recent years have revealed risk factors, which describe the women most likely to develop ovarian cancers (Beral et al, 1972, Norris and Jensen, 1972, Berg and Baylor, 1973, Raymond and Bonenfant, 1976, Editorial-2, 1978, Muir and Nectoux, 1978). The more important of these risk factors are: unmarried status, hormonal abnormalities (infertility, few pregnancies, menstrual disorders), ovarian dysgenesis, familial and genetic traits, race and environment (exposure to carcinogens), and age. Although several interesting studies report the occurrence of ovarian adenocarcinomas in families (Lynch et al, 1972, Li et al 1970), suggesting gene defects, most authorities agree environment is more important (Berg, 1975). When we analyze these risk factors more closely we find almost all can be correlated with hormonal abnormalities and-or reproductive handicaps, perhaps due to gonadal defects. Thus we find some experts suggesting over-nutrition (Editorial-1, 1977), that is, high-fat, high-protein diets which act early in life and predispose to cancer by possibly influencing prolactin levels. Other investigators postulate low dietary iodine (Stadel, 1976), a deficiency that might stimulate gonadotropin secretion, which in turn would induce hyperestrogenism.

That environmental factors are highly important is obvious from the observation that Seventh Day Adventists and Mormons in America experience a lower risk for ovarian cancer than do other women in the United States (Lyon et al, 1976).

In examining the morphology of cystadenofibromas of the ovary, Papadaki and Beilby (1975) believed they could from histological criteria alone implicate the role of estrogenic hormones in the pathogenesis of these tumors. Recently Hoover and associates (1977, 1978) reported a 30-fold risk for ovarian carcinoma in women who received diethylstilbestrol during a pregnancy; the risk for estrogens was 2 to 3 fold. Beral and coworkers (1978) considered that the rise in mortality from ovarian carcinoma in England and Wales was best explained by the decrease in childbearing of the last century, supporting the idea that pregnancy affords a rest and protection for the ovary.

Fathalla (1971, 1972) went a step further to postulate that ovarian adenocarcinoma is related to the incessant purposeless ovulations which unmarried women or nulliparous women experience. The incessant ovulations, he suggested, provided more chances for germinal inclusions and cysts to develop, which he thought could be precursors of the carcinomas. In their studies of ovarian cancer in Quebec, Raymond and Bonenfant (1976) reported 51 women with ovarian carcinomas, all of whom had born 10 or more children, suggesting multiparity is no sure safeguard against developing such tumors.

Methods used to produce ovarian tumors experimentally give us further leads about how they might arise in women (Lipschutz et al, 1967, Henderson et al, 1971, Brush, 1973, Lingeman, 1974, Woodruff, 1975). The common denominator of most is a hypersecretion of gonadotropins which induces the ovarian tumors (Stadel, 1975).

Xrays and carcinogens, for example, destroy the estrogen producing cells of the ovarian follicles, so that the fall in plasma estrogen stimulates gonadotropin secretion. In many species of animals, however, ovarian tumors can be produced by injections of hormones such as diethylstilbestrol in dogs (Jabara 1962) or synthetic progestogens in mice (Lipschutz et al, 1967, Myhre, 1972). These exogenous hormones act by way of feedback mechanisms on the hypothalamus and pituitary stimulating gonadotropins which induce the tumors.

From the ease with which polycyclic aromatic hydrocarbons induce ovarian tumors in rodents (Howell and Orr, 1954), some investigators have suggested similar mechanisms may act in women. Statistics show, for example, that women working in rubber, electrical, or textile industries are at greater risk for ovarian carcinoma than are women in other industries, explaining too why ovarian cancer occurs more commonly in industrial countries.

Recently Mattison and Thorgeirsson (1978) defined enzyme systems in mouse ovaries that enable the ovarian cells to metabolize aromatic hydrocarbons to intermediates which can cause cancer. These workers have postulated that aromatic hydrocarbons which women smokers might absorb could react in their ovaries in like manner. Although Berg and Baylor (1973) declared, "we know from many studies that cigarettes are not epidemiologically linked with endocrine-dependent cancers", recent reports provide evidence that women smokers experience a menopause earlier than expected because they lose estrogenproducing follicles prematurely (Jick et al, 1976).

Since the methods used to produce ovarian tumors in animals involve destruction of follicular epithelium or prevention of ovulation (x-irradiation, carcinogens, hormonal imbalance, ligature of ovarian vessels), the tumors that arise are histologically and biologically not comparable with the adenocarcinomas developing in older women. These cancers in women, according to the major theory, arise from cysts formed from either dormant remnants of müllerian epithelium or from snipped-off inclusions of germinal epithelium caught up in the ovary after an ovulation (Fathalla, 1971, Griffiths et al, 1973). In either case, these epithelial cells are multipotential and responsive to hormones, undergoing mitoses when exposed to estrogens. If adenocarcinomas do develop from cystomas and these from germinal epithelium caught up after ovulation, then by reducing ovulations one might decrease the incidence of ovarian cancers, as some have already postulated (Fathalla, 1971, Ory, 1974, Zajicek, 1978). To test that postulate we asked the computer, which contained the tumor registry of the Mannheim Women's Hospital, to tell us how many of the women from whom ovarian tumors had been removed had taken oral contraceptives. Of 80 epithelial tumors, ten were from women who had taken the pill. Cystadenomas, as our statistics indicate, are becoming more common despite the wide use of oral contraceptives.

Because breast, endometrial and ovarian cancers often occur in the same patient, one is tempted to contrive a single basic causal mechanism to explain their origins.

Since ovarian carcinoma is becoming a common disease and as many studies indicate, is related to hormonal imbalances and possibly fat-soluble carcinogens as well, it behooves us to learn more about its pathogeneses as soon as possible. If hormonal imbalances reflect changes in childbearing patterns or therapy with sex hormones, then the only way we can pinpoint these facts is by close cooperative studies.

References

Beral V, Fraser P, Chilvers C (1978) Does pregnancy protect against ovarian cancer? Lancet *I*, 1083–1087

Berg JW (1975) Can nutrition explain the pattern of international epidemiology of hormone-dependent cancers? Cancer Res. *35*, 3345–3350

Berg JW, Baylor SM (1973) The epidemiologic pathology of ovarian cancer. Hum. Path. *4*, 537–547

Brush, MG (1973) Experimental approaches to endometrial and ovarian cancer. Postgrad. Med. Jour. *49*, 73–76

Editorial-1 (1977) Aetiological factors in ovarian cancer. Lancet *II*, 1062–1063

Editorial-2 (1978) Epidemiology of ovarian cancer. Brit. Med. Jour. *1*, 198

Fathalla MF (1971) Incessant ovulation – a factor in ovarian neoplasia? Lancet *II*, 163

Fathalla MF (1972) Factors in the causation and incidence of ovarian cancer. Obstet. Gynec. Survey *27*, 751–768

Griffiths K, Henderson WJ, Chandler JA, Joslin CAF (1973) Ovarian cancer: some new analytical approaches. Postgrad. Med. J. *49*, 69–72

Henderson WJ, Joslin CAF, Turnbull AC, Griffiths K (1971) Talc and carcinoma of the ovary and cervix. Jour. Obstet. Gynecol. (Brit. Commw.) *78*, 266–272

Hoover R, Gray L, Fraumeni J (1977) Stilbestrol and ovarian cancer. Lancet *II*, 533

Hoover R, Fraumeni JF, Gray LA (1978) Exogenous oestrogens and ovarian cancer. Lancet *I*, 325

Howell JS, Orr JW (1954) The induction of ovarian tumours in mice with 9: 10 dimethyl-1:2 benzanthracene. Brit. J. Cancer *8*, 635–646

Jabara AG (1962) Induction of canine ovarian tumours by diethylstilbesterol and progesterone. Austral. J. Exp. Biol. *40*, 139–195

Jick H, Porter J, Morrison AS (1976) Relation between smoking and age of natural menopause. Lancet *I*, 1354–1355

Li FP, Rapoport AH, Fraumeni JF, Jensen RD (1970) Familian ovarian carcinoma. JAMA *214*, 1559–1561

Lingeman CH (1974) Etiology of cancer of the human ovary: a review. J. Nat. Can. Inst. *53*, 1603–1618

Lipschutz A, Iglesias R, Panasevich VI, Salinas S: Ovarian tumours and other ovarian changes induced in mice by two 19-nor-contraceptives. Brit. J. Cancer *21*, 153–159

Lynch HT, Krush AJ, Lemon HM, Kaplan AR, Condit PT, Bottomley RH (1972) Tumor variation in families with breast cancer. JAMA *222*, 1631–1635

Lyon JL, Klauber MR, Gardner MS, Smart CR (1976) Cancer incidence in mormons and non-mormons in Utah, 1966–1970. N. Engl. J. Med. *294*, 129–133

Mattison DR, Thorgeirsson SS (1978) Smoking and industrial pollution, and their effects on menopause and ovarian cancer. Lancet *I*, 187–188

Muir CS, Nectoux J (1978) Ovarian cancer: some epidemiological features. World Health Stat. Rep. *31*, 51–61

Myhre E (1972) Ovarian morphology following long-term treatment with sex hormones and contraceptive steroids in mice. Acta path. microbiol. scand. Sect A. *80*: suppl. 233: 67–75

Norris HJ, Jensen RD (1972) Relative frequency of ovarian neoplasms in children and adolescents. Cancer *30*, 713–719

Ory H (1974) Functional ovarian cysts and oral contraceptives negative association confirmed surgically. JAMA *228*, 68–69

Papadaki L, Beilby JOW (1975) Ovarian cystadenofibroma: a consideration of the role of estrogen in its pathogenesis. Amer. J. Obstet. Gynecol. *121*, 501–512

Raymond H, Bonenfant J (1976) Epidemiology of ovarian cancer. L'Union Med. du Canada *105*, 1048–1053

Stadel BV (1975) The etiology and prevention of ovarian cancer. Am. J. Obstet. Gynecol. *123*, 772–774

Stadel BV (1976) Dietary iodine and risk of breast, endometrial and ovarian cancer. Lancet *I*, 890–891

Woodruff JD: Experimental production of ovarian neoplasia. In: De Watteville, H.Ed. Diagnosis and Treatment of Ovarian Neoplastic Alterations. Amsterdam, Excerpta Medica, 1975. pp 53–55

Zajicek J (1978) Prevention of ovarian cystomas by inhibition of ovulation: a new concept. Jour. Reprod. Med. 20, 114

Histologic Changes of the Cervix Uteri Following Contraceptive Use

H. Schaude, Gisela Dallenbach-Hellweg and K. Schlaefer

Exogenous sex hormones are known to induce changes in the cervix uteri, an effect we have become particularly aware of since oral contraceptives came into use. As early as 1961 reports began to appear that ectopies were found more frequently in women who had been on hormone therapy than in women who had not (Cook et al. 1961; Pincus 1965; Kehrer and Hauser 1968; see Table 1). The most pronounced ectopies were linked with drugs of high gestagen potency. Dallenbach-Hellweg (1972) traced that phenomenon to the stimulus to proliferation that exogenous hormones exert in the endocervix. Estrogens, by contrast, seem to have a more differentiating effect, particularly on the squamous epithelium of portio uteri and vagina. The most common histologic signs of the proliferative stimulus of progesterone are adenomatous hyperplasia of the cervix mucosa and reserve cell hyperplasia.

We have conducted statistical analyses of histologic material from a large group of patients, and found that ectopies and dysplasias were significantly more frequent in women who had taken oral contraceptives than in those who had not (Dallenbach-Hellweg et al. 1971a, b, c; Mohr 1973). A number of other authors, whose results were based on cytologic studies, found similar frequency distributions (see Table 2, and for a discussion, Dallenbach-Hellweg 1972).

Material and Method

We have attempted to determine the connection between morphological changes in the cervix uteri and exogenous hormones by analyzing all the case reports we received

Table 1. Frequency of cervical ectopies

Author	Date	Not on contraceptives	On contraceptives	Type
Cook et al.	1961	49%	70%	Norethynodrel
Pincus	1965	46%	63%–71%	Enovid
Kehrer and Hauser	1968	Total: 39,4% 1st Group: 0,0%	79,9% 70%	Anovlar, Lyndiol, Noracyclin
Mohr	1973	34%	60%	

Table 2. Twice normal frequency of cervical dysplasia after contraceptive use in women under 40 was found by following authors

Author	Date	Women investigated	
		Total number	Numer using contraceptives
Attwood	1966	9,500	500
Melamed et al.	1969	39,954	27,508
Stern et al.	1970	1,614	1,476
Kline et al.	1970	20,020	17,724
Dougherty	1970		1,983
Kirkland u. Stanley	1970	146,565	26,286
Bibbo et al.	1971	148,735	18,380
Powell u. Seymour	1971		1,107
Dallenbach-Hellweg	1972	1,096	375
Mohr	1973	38,000	
Total		405,484	95,339

between 1975 and 1978. To do that, we programmed the personal data, clinical histories, and histologic diagnoses of 20,338 patients and fed this information into an IBM 370–158 computer with a working storage capacity of three megabytes. A long-distance line was set up for the purpose between the Department of Gynecology and Obstetrics Mannheim and the Heidelberg Cancer Research Center.

Data was corrected at the terminal, and evaluated by cross-tabulating with the Statistical Package for the Social Sciences. We also used some of the programs of the INDA data bank system.

Since despite our repeated requests we were not able to obtain complete case histories of hormone use from some of the patients of individual doctors, we selected our data for evaluation in three stages. First, we selected only those physicians who had

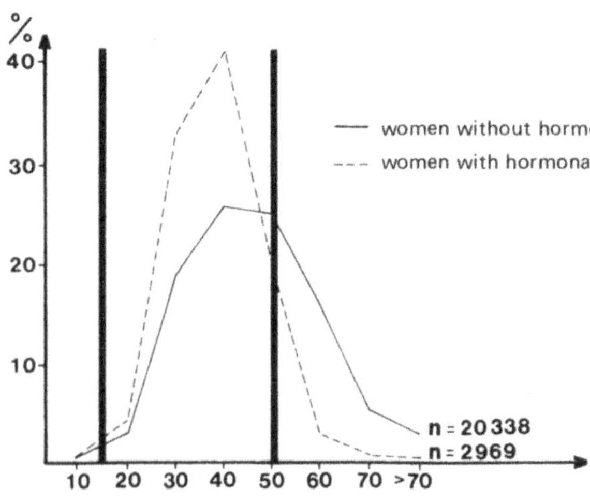

Fig. 1. Age distribution. The group of women using contraceptives is younger than the control group, with a peak between 25 and 45 years of age.

submitted complete clinical histories, and second, considered their information as reliable only if at least 30% of their patients had been receiving contraceptives. We chose that percentage arbitrarily, because we believed it probably best reflected the real percentage for the population as a whole. Third, we limited our material by evaluating only data from women between the ages of 15 and 50 years.

The age distribution of the collective as a whole conformed approximately to the normal curve. The same was true for the group of women on exogenous hormones, whose age curve was shifted to the left, with a peak between the ages of 25 and 45, a time, of course, when family planning is most opportune. Our group of women on contraceptives was thus younger on the average than the nonmedicated group (Fig. 1). By taking the cutoff age at 50 we corrected for that age difference. In the 4-year period from 1975 to 1978, histologic examinations had been performed in a total of 20,338 women, resulting in about 40,000 diagnoses.

Preliminary Results of the Evaluation

After limiting our data by the procedure described above, we obtained a *first collective* of 1,774 diagnoses of cervical epithelium. The type of operation used for diagnosis was considered irrelevant, whether conisation, excision, curettage or hysterectomy. Of the diagnoses, 387 were made in women who had not been taking oral contraceptives and 1387 in women who had (Table 3).

As regards benign histologic changes, adenomatous hyperplasia of cervical mucosa was four times more frequent in the contraceptive group than in the noncontraceptive group. Reserve cell hyperplasia was also significantly increased by exogenous hormones. By contrast, squamous cell metaplasia was found more frequently in women who had taken no contraceptives (Table 4). On the whole, alterations in the cervix epithelium were diagnosed more frequently in connection with hormone use than without.

Our evaluation of preneoplastic lesions of the cervix was based on the number of uterine tissue specimens taken from patients selected by the above criteria over the 4-year period. Of the total of 629 specimens taken by conisation or hysterectomy, 186 were from women who had not used, and 443 from women who had used oral contracepti-

Table 3. Frequency of cervical changes in women with and without hormonal medication in a selected sample

Benign cervical changes		
Epithelial diagnoses in women from 15 to 50 years of age (data submitted by 9 selected physicians)		1774
Of these, without hormonal medication	387	
And with hormonal medication	1387	
		1774

Table 4. Benign changes in cervical epithelium (1774 diagnoses)

	Without hormones (n = 387)		With exog. hormones (n = 1387)	
Adenomatous hyperplasia	24	6,2%	366	26,4%
Reserve cell hyperplasia	42	10,9%	201	14,5%
Squamous cell metaplasia	103	26,6%	245	17,6%
	169	41,7%	812	58,5%

Table 5. Benign and preneoplastic changes in the cervix (629 cervical specimens)

	without hormones n = 186	%	with exogenous hormones n = 443	%
Adenomatous hyperplasia	24	12,9	366	82,6
Reserve cell hyperplasia	42	22,6	201	45,4
Squamos cell metaplasia	103	55,4	245	55,3
Dysplasia	28	15,1	106	23,9
Carcinoma in situ	3	1,6	14	3,2

ves. We found a significantly higher frequency of dysplasia and carcinoma in situ in the exogenous hormone group as against the controls (Table 5).

Our *second collective* for evaluation comprised 13,687 women between the ages of 15 and 50 who had seen a gynecologist and from whom tissue of some kind (ovary, tube, uterus, vagina, vulva, or breast) had been taken for histologic examination. The percentages of benign, preneoplastic and neoplastic alterations in the cervix uteri were correspondingly much lower in this collective, yet the differences between pill-users and non-pill-users were comparable all the same (Table 6). Here the larger number of patients made it obvious that invasive carcinomas were not increased by contraceptives, but that the precancerous stages appeared almost twice as frequently in contraceptive users than in controls. An evaluation of this collective by age revealed that ectopies, adenomatous hyperplasias, dysplasias, and carcinoma in situ all occurred at a relatively earlier age in contraceptive users.

The *third collective* comprised 2,542 cervical ectopies and the preneoplasias that develop as a result (Table 7). Disorders of that type were also found to occur slightly ear-

Table 6. Women between the ages of 15 and 50 from whom tissue was taken (from ovary, fallopian tube, uterus, vagina, vulva, or breast) for histologic examination (Total no.: 13,687)

	Without contraceptives				Totals	With contraceptives				Totals
Age in years	15–20	21–30	31–40	41–50		15–20	21–30	31–40	41–50	
Totals	305	1986	3024	3538	8853	198	1530	1874	1232	4834
Infl. gland. papill. ectopia	7	142	528	695	1372	34	389	482	265	1170
%	2.3	7.2	17.5	19.6	15.5	17.2	25.4	25.7	21.5	24.2
Dysplasia	3	23	60	109	195	7	65	98	57	227
%	1.0	1.2	2.0	3.1	2.2	3.5	4.3	5.2	4.7	4.7
Cacinoma in situ	0	5	11	18	34	1	17	27	13	58
%	0.0	0.3	0.4	0.5	0.4	0.5	1.1	1.4	1.1	1.2
Microcarcinoma	0	1	5	7	13	0	2	9	2	13
%	0.0	0.1	0.2	0.2	0.1	0.0	0.1	0.5	0.2	0.3
Invasive carcinoma	0	1	1	9	14	0	3	4	1	9
%	0.0	0.1	0.0	0.34	0.2	0.0	0.2	0.2	0.2	0.2

Table 7. Preneoplastic and neoplastic changes in the portio uteri (2,542 portio ectopies)

	Without contraceptives				Totals	With contraceptives				Totals
Age in years	15–20	21–30	31–40	41–50		15–20	21–30	31–40	41–50	
Totals	7	142	528	695	1372	34	389	482	265	1170
Dysplasia	3	23	60	109	195	7	65	98	57	227
%	–	16.2	11.4	15.7	14.2	20.6	16.7	20.3	21.5	19.4
Carcinoma in situ	0	5	11	18	34	1	17	27	13	58
%	–	3.5	2.1	2.6	2.5	2.9	4.4	5.6	4.9	5.0
Microcarcinoma	0	1	5	7	13	0	2	9	2	13
%	–	0.7	0.9	1.0	0.9	–	0.5	1.9	0.8	1.1
Invasive carcinoma	0	1	1	9	11	0	3	4	1	8
%	–	0.7	0.2	1.3	0.8	–	0.8	0.8	0.4	0.7

lier in women who had been taking contraceptives, and the frequency of dysplasia and carcinoma in situ was one and a half to two times higher in this group than in the controls. Invasive carcinoma, as in the other collectives, was found to occur in a relatively low percentage of cases, irrespective of contraceptive use.

Interpretation of Findings

The high frequency of adenomatous hyperplasias and reserve cell hyperplasias observed in the contraceptive groups points to the proliferative effect that exogenous hormones, and particularly synthetic gestagens, have on the endocervical mucosa. By contrast, without hormones, the differentiating components predominate, which means that regenerative processes such as reserve cell hyperplasia can mature into squamous cell metaplasia. All three of the collectives studied so far were remarkable for the high frequency with which preneoplastic epithelial lesions were associated with exogenous hormones, an effect already described elsewhere in the literature (Dallenbach-Hellweg et al. 1971a, b, c; Mohr 1973; Stern et al. 1977).

Since the number of invasive carcinomas did not increase with contraceptive use, we believe that means women who take the pill usually undergo earlier and more frequent gynecologic examinations. It also underscores the importance of cancer tests, which, from the results of our investigation, we think are absolutely justified. We do not believe that there is a direct connection between exogenous hormones and neoplastic changes in the cervix. However, it has been shown that contraceptives, particularly those high in gestagen, can lead indirectly, via chronic inflammation or increased proliferation, to the development of precancerous lesions such as dysplasia and carcinoma in situ.

The figures given in this paper should be considered preliminary[1]. We are planning to analyze the extensive data further by such factors as type of tissue removed, number of births and pregnancies, infections, and correlate these with other secondary findings. As far as possible we also hope to take social factors into consideration. Since, however, the use of contraceptives is associated with constant changes in the mucosa of portio and cervix, changes that differ in absolutely characteristic ways from normal histologic structure, we do not believe that social factors will prove of more than secondary significance in their causal genesis. Of interest in this connection is a group of young patients who were prescribed contraceptives for purely medical reasons (e.g., as treatment for acne). These young women developed ectopies, adenomatous hyperplasias and dysplasias to the same extent as our other groups even though they had never had sexual intercourse and were all virgo intacta. Though this group is as yet very small, we are continuing to study it.

1 While this paper was being prepared for publication, the data were analyzed in detail and proved to be of high statistical significance

References

Attwood ME (1966) Cytology and the contraceptive pill. J Obstet Gynec Brit Cwlth 73:662

Bibbo M, Keebler CT, Wied GL (1971) Prevalence and incidence rates of cervical atypia. A computerized file analysis on 148735 patients. J Reproduct Med 6:184

Cook HH, Gamble CJ, Satterwaite AP (1961) Oral contraception by norethynodrel. Am J Obstet Gynecol 82:437

Dallenbach-Hellweg G (1972) Therapieschäden in der Gynäkologie . Verh Dtsch Ges Pathol 56:252

Dallenbach-Hellweg G, Herting W, Momber F, Thorn V (1971a) Portioveränderungen unter Ovulationshemmern. Dtsch Ärztebl 68:2288

Dallenbach-Hellweg G, Herting W, Momber F, Thorn V (1971b) Portioveränderungen unter Ovulationshemmern. Verh Dtsch Ges Pathol 55:682

Dallenbach-Hellweg G, Herting W, Momber F, Thorn V (1971c) Zytologische und histologische Untersuchungen der Portio und Cervix uteri unter der Einnahme von Ovulationshemmern. Fortschr Med 89:883

Dougherty CM (1970) Cervical cytology and sequential birth control pills. Obstet Gynec 36:741

Kehrer, B, Hauser GA (1968) Häufigkeit und Auftreten der Portioektopien unter Ovulationshemmern. Gynaecologia 165:209

Kirkland KA, Stanley MA (1970) Oral contraceptives and cervical neoplasia. Cancer Cytology 10:9

Kline TS, Holland M, Wemple D (1970) Atypical cytology with contraceptive hormone medication. Am J Clin Pathol 53:215

Melamed MR, Koss LG, Flehinger BJ, Kelisky RP, Dubrow H (1969) Prevalence rates of uterine cervical carcinoma in situ for women using the diaphragm or contraceptive oral steroids. Brit Med J 3:195

Mohr HJ (1973) Epithelveränderungen der Portio- und Zervixschleimhaut ohne und mit Ovulationshemmer-Einwirkung. Fortschr Med 91:578

Pincus G (1965) Control of fertility. Academic Press New York

Powell LG, Seymour RJ (1971) Effects of depomedroxyprogesterone acetate as a contraceptive agent. Am J Obstet Gynecol 110:36

Stern E, Clark VA, Coffelt CF (1970) Contraceptives and dysplasia: higher rate for pill choosers. Science 169:497

Stern E, Forsythe AB, Youkeles L, Coffelt CF (1977) Steroid contraceptive use and cervical dysplasia: Increased risk of progression. Science 196:1460

Comparative Study of Vaginal Cytology Involving Controls and Patients Receiving Oral Contraceptive Agents

H. Joswig-Priewe and K. Schlüter

Abstract

We carried out a retrospective cytologic study comparing a group of 692 patients using oral contraceptives with a group of 309 patients receiving no hormonal therapy. Statistical means were applied to select the two groups, and the cytological smears were evaluated without knowledge of source.

1. From morpholigical criteria we were able to assign 75% of the smears to the correct group.
2. It proved possible from special morphological changes to differentiate the effect of estrogen-dominant, gestagen-dominant or, hormonally balanced oral contraceptives.
3. The women using oral contraceptives showed a 12,9% higher rate of inflammation, and a 8,1% higher rate of inflammatory changes of no detectable cytologic or clinical cause.
4. In 2,8% (20 patients) who took the oral contraceptives for 6 months to six years we found atrophic cell pictures.
5. For those taking the pill, the rate for dysplasia was 1,25%, compared with 0,97% for those receiving no hormonal therapy.

The Effect of Combination Contraceptive Steroids on the Mammary Gland

H.J. Norris

The effects of reproductive steroids on animal mammary glands should not be compared with those on the human breast because animals differ from humans in their endogenous steroids, estrus patterns, gestations, and lactational histories. There are also marked differences in life span, genetic influences, mammary histology, and the types of spontaneous neoplasms. Beagle dogs seem especially prone to mammary nodules and are therefore a poor test animal for comparison with the human (Nelson et al. 1973). Since there is so little basis for comparison, only the effect of combined oral contraceptive steroids on the mammary tissues on primates is considered below. The effects of oral contraceptive steroids on the mammary glands of various animal species are summarized elsewhere (Casey et al. 1979; Hamilton 1974).

Nonhuman Primates

Very few nonhuman primates have been observed clinically throughout a normal life span and only 12 mammary tumors have been reported, 8 of which were carcinomas (Casey et al. 1979). One carcinoma occurred in a group of six monkeys given normal human doses of an oral contraceptive combination for 18 months (Kirschstein et al. 1972). In one study of the long-term effect of synthetic steroid combinations, 96 monkeys received 1, 10, and 50 times the recommended human dose, yet none developed palpable tumors during 5 years of observation (Drill et al. 1974). In a similar study, however, 23 monkeys receiving synthetic steroid combinations developed transitory nodules (Geil and Lamar 1977).

Intraductal hyperplasia has been identified in biopsies of clinically normal mammary glands of monkeys treated with a variety of synthetic estrogens and progestins (ethynerone, chloroethynyl, norgesterel, anagestone, mestranol, and medroxyprogesterone acetate) (Wazeter et al. 1976). However, the intraductal hyperplasia seen in test monkeys has not yet been compared with possible changes in control animals. Further clinical observation followed by thorough histologic examination is needed to determine the biologic significance of the hyperplasia in monkeys.

Humans

The human breast is under the influence of endogenous ovarian steroids during the menstrual cycle as well as prolactin and growth hormone synthesized and released by the pituitary gland. Insulin, thyroxin, and cortisol must also be available for full differentiation of the mammary gland. The growth of ductal epithelium and maintenance of the stroma of the breast is largely via estrogen, while progesterone promotes growth and development of lobules and acini and induces secretory activity.

Morphological Considerations

The histologic findings ascribed to oral contraceptive steroids in the breast of women are variable, but include an increase in the size of lobules from dilation and proliferation of acini, enlargement of acinar cells by increased secretory activity, and edema of the intralobular stroma. These changes vary from one area of the breast to another and may be diffuse, focal, or absent (Fechner 1977). The findings are not specific for oral contraceptive users since similar, but more advanced changes, occur in lactating and pregnant women. Mammography has found increased density of breasts while women take oral contraceptives, but there are no localizing signs.

Oral contraceptives are prescribed to patients having symptoms from fibrocystic disease, with subjective improvement in some instances. Epithelial hyperplasia and atypia have not been more frequent in oral contraceptive users (LiVolsi et al. 1978). Intraductal hyperplasia has been reported in fibroadenomas taken from women using oral contraceptives (Brown 1970; Goldenberg et al. 1968; Prechtel and Seidel 1973), but in studies of 54 biopsies containing fibroadenomas and 25 biopsies with fibrocystic disease from women who were taking an oral contraceptive, no distinctive features were found when compared with biopsies from age-matched control women who had not used oral contraceptives (Fechner 1970a, b). Galactorrhea occurs in some patients taking oral contraceptives, particularly those with higher doses of estrogen. A variety of drugs affect lactation, most of them through suppression of hypothalamic prolactin-inhibiting factor (Dickey and Stone 1975). Normally, estrogen used alone tends to suppress lactation, whereas progesterone, used alone, does not produce galactorrhea, even though it does stimulate secretion (Dickey and Stone 1975).

Women with mammary cancers have not reported an unusually high frequency of oral contraceptive use when compared with control subjects. Nonetheless, carcinomas of the breast having distinctive features have been reported in women taking oral contraceptives (Gould et al. 1972; Penman 1970). Both the carcinomas and the adjoining breast tissue showed the effects of stimulation by estrogen and progesterone as the intralobular stroma was edematous, myxoid and basophilic, and infiltrated in areas by lymphocytes and plasma cells (Figs. 1 and 2). The acinar cells were eosinophilic and granular with a marked secretory affect, and "colostrum" cells were present in one case (Penman 1970). Some lobules in each case were enlarged (Fig. 3) and distended by secretion. These features are similar to those within some carcinomas excised from patients who are lactating or who are pregnant and near term (Norris and Taylor 1970). Carcinomas from women taking oral contraceptives usually show no differences from nonpregnant women of the same age, but the myxoid periductal and intralobular

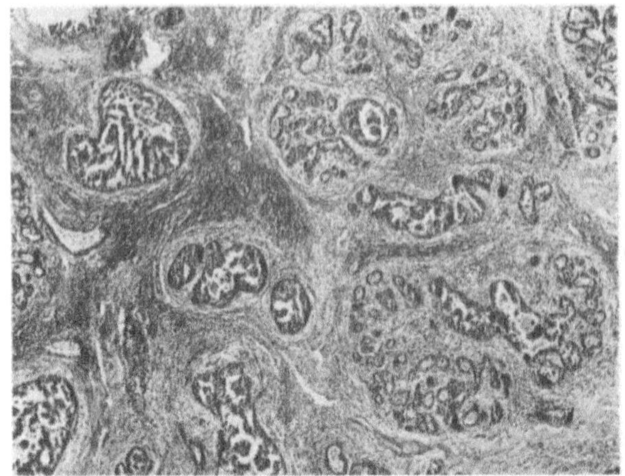

Fig. 1. Biopsy from the breast of a patient taking an oral contraceptive steroid. Lobules *(right)* and ducts *(left)* show markedly atypical alterations. Prominent lymphoid stroma is also present. H and E stain, AFIP Neg. 77-5619. x 28.7

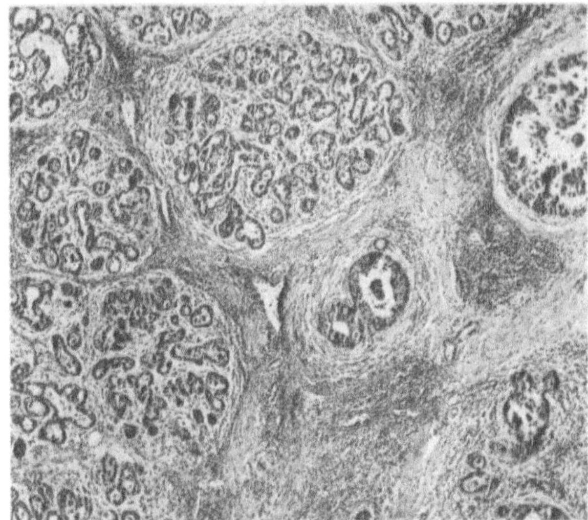

Fig. 2. Same biopsy as Fig. 1 showing lobules lined by atypical cells and dilated acini containing a secretory product, presumably a steroid effect. H and E stain, AFIP Neg. 77-5591. x 46

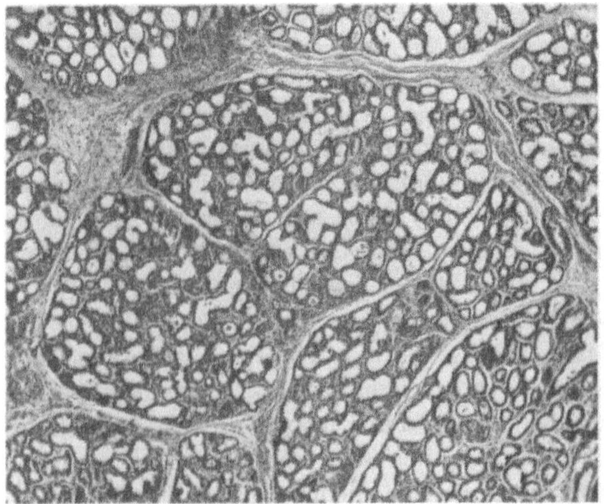

Fig. 3. Same breast biopsy as shown in Fig. 1 and 2. Note marked lactational effect in the otherwise normal lobule. This degree of lobular hyperplasia is not seen in the absence of hormone stimulation by nursing, pregnancy, oral contraceptives, or certain nonsteroidal galactogenic central nervous system stimulants. H and E stain, AFIP Neg. 77-5594. x 37.4

change (Figs. 4 and 5) coupled with transformation of lobules by a glandular adenocarcinoma (Fig. 6) is very suggestive of steroid use. If an unusual pattern appears repeatedly such as the lipid rich carcinoma shown in Fig. 7, this becomes evidence that oral contraceptives are modifying or inducing carcinomas of the breast. In most reports, the oral contraceptives did not induce the carcinomas because the duration of

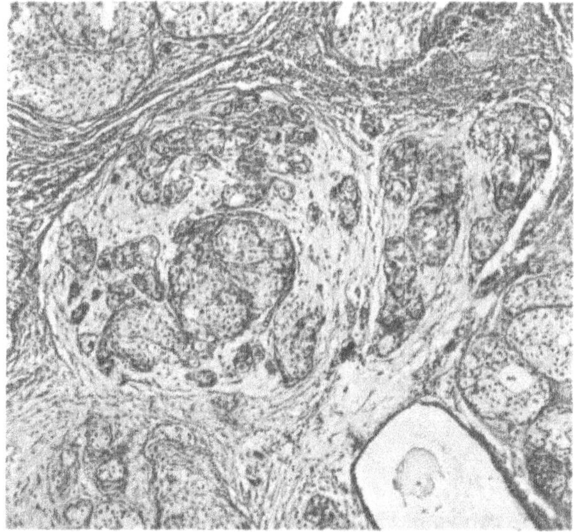

Fig. 4. This carcinoma from the breast of a patient taking an oral contraceptive shows a prominent myxoid stroma, an estrogen effect. H and E stain, AFIP Neg. 77-5727. x 56.2

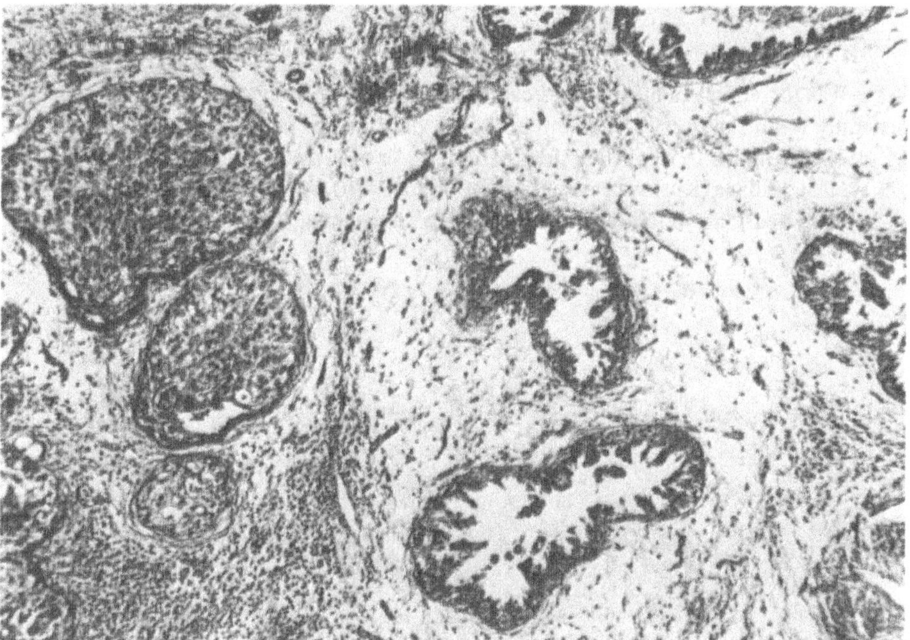

Fig. 5. Intraductal carcinoma with marked edema and myxoid alteration of the periductal stroma from a patient taking an oral contraceptive. H and E stain, AFIP Neg. 70-78-24. x 58.4

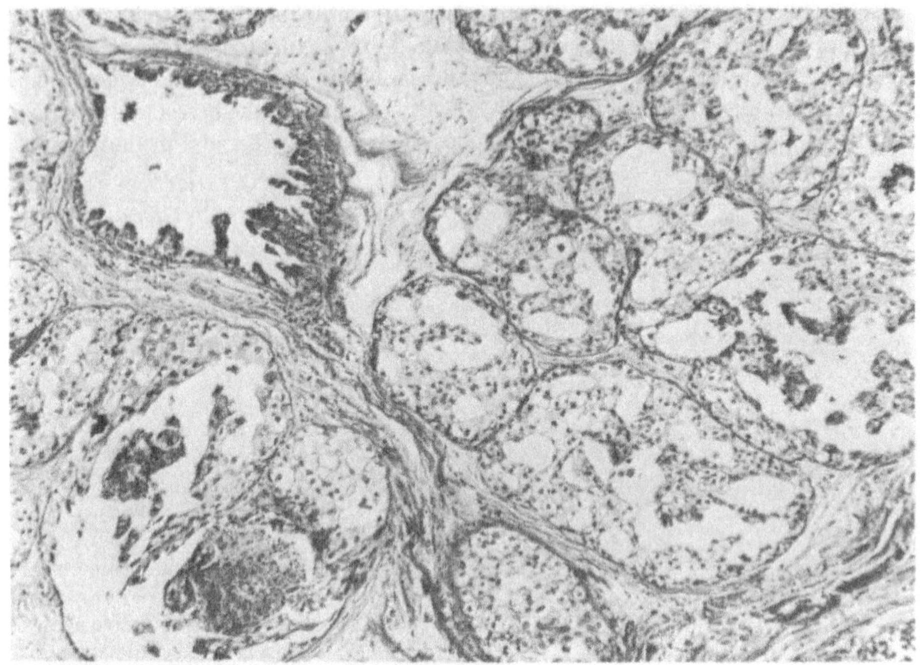

Fig. 6. Elsewhere in the same breast as shown in Fig. 5 there is a prominent glandular component in an exaggerated lobular pattern. H and E stain, AFIP Neg. 70-7825. x 58.4

usage was too short. Induction of a cancer by a chemical carcinogen is a long process, generally requiring years before the initial neoplastic transformation occurs (Kusama et al. 1972; Pearlman 1976). Also, the doubling time of mammary cancer growth averages at least 25 days (Kusama et al. 1972; Perlman 1976). With that rapid doubling time, it takes 2 years after induction before the average neoplasm becomes clinically

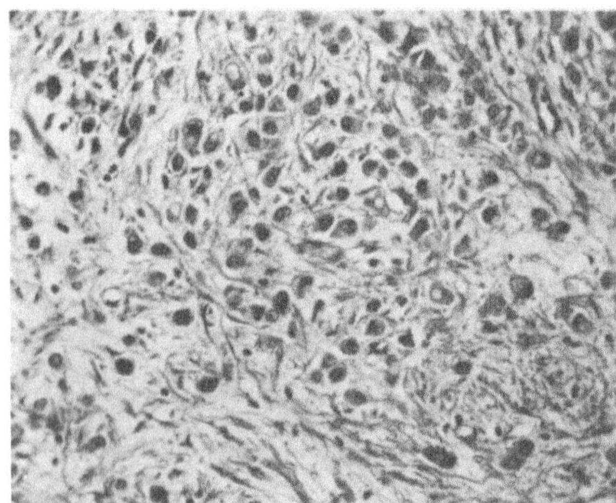

Fig. 7. This rare type of carcinoma is a "lipid rich" carcinoma resembling histiocytic or signet cell carcinoma. The patient, 35 years old, had been taking an oral contraceptive for many years. H and E stain, AFIP Neg. 77-5219. x 176.8

evident. Most investigators have found much slower doubling times for the growth of mammary carcinoma, the average being around 100 days, so that it takes an estimated 9 years after inception for a neoplasm to become clinically evident. For those reasons, the unusual features reported in carcinomas from women taking oral contraceptives reflect modification of a preexisting carcinoma and the adjoining noninvolved mammary tissue by steroids. Nonetheless, the reports serve as a warning that if oral contraceptives can modify the histologic appearance of mammary carcinomas in rare instances, then they could also alter the growth rate and perhaps serve as a cocarcinogen in rare instances.

Contraceptive steroids have been utilized in therapy of clinical fibrocystic disease (Arial 1973), but it is not clear whether they have any value. Some patients report tenderness, fullness, and even an increase in the size of their breasts while taking oral contraceptives. When these symptoms occur, it is with the initial use, and after several cycles most will subside. Occasionally, because of symptoms there may be a need to either switch preparations or discontinue their use altogether.

Epidemiologic Considerations

Despite the worrisome morphological findings in human and nonhuman primates, the epidemiologic data is comforting in that there is no evidence at present that oral contraceptives induce mammary cancer or atypical hyperplasia in women. In fact, there is good evidence that oral contraceptives reduce the risk of benign diseases of the breast. At least seven case control studies have been undertaken (Table 1) with six of the seven showing that the relative risk for a variety of types of benign diseases such as fibrocystic disease and fibroadenomas is less in users of oral contraceptives than controls (LiVolsi et al. 1978; Ravnihar et al., cit. by Thomas 1978; Kelsey et al. 1978; Sartwell et al. 1973; Vessey et al. 1972; Paffenbarger et al. 1977; Boston Collaborati-

Table 1. Case-control studies of the relationship of oral contraceptives to risk of benign diseases of breast[a]

First author	No. of cases	Relative risk	Minimum years of use
Ravnihar et al., cit. by Thomas 1978	419	0.48	2
Kelsey et al. 1978	366	0.35	5
Sartwell et al. 1973	1048	0.69	5
Vessey et al. 1972	255	0.47	6
Paffenbarger et al. 1977	446	0.20	8
BCDSP[b]	98	0.47	–
Nomura and Comstock 1976	318	1.14	–

[a] Modified from Thomas (1978)
[b] Boston Collaborative Drug Surveillance Programme (1973)

Table 2. Case-control studies of the relationship of oral contraceptives to the risk of breast cancer[a]

First author	No. of cases	Relative risk for long-term users	Minimum years of use
Vessey et al. 1972	322	0.94	2
BCDSP[b]	23	0.60	–
Paffenbarger et al. 1977	452	1.7	8
Henderson et al. 1974	308	0.74	–
Sartwell et al. 1977	284	0.97	5
Kelsey et al. 1978	99	1.7	5
Ravnihar et al., cit. by Thomas 1978	190	0.86	2

[a] Modified from Thomas (1978)
[b] Boston Collaborative Drug Surveillance Programme (1973)

ve Drug Surveillance Programme 1973; Nomura and Comstock 1976). In fact, some suggest a dimunition in risk with increasing duration of use of oral contraceptives. The "protective effect" of oral contraceptives appears to be particularly strong for cystic disease of the breast (Thomas 1978).

A prospective study by the Royal College of General Practitioners (1977) suggested that the rates of benign breast disease are inversely related to the amount of progestin in the contraceptive.

At least seven case control studies of patients with breast cancer and the use of oral contraceptives have been conducted (Ravnihar et al., cit. by Thomas 1978; Kelsey et al. 1978; Vessey et al. 1972; Paffenbarger et al. 1977; Boston Collaborative Drug Surveillence Programme 1973; Henderson et al. 1974; Sartwell et al. 1977). None of these show a significant increase in mammary carcinoma in patients who have ever used oral contraceptives (Table 2). The epidemiologic aspects of the case control studies are critically analyzed elsewhere (Thomas 1978).

There have been three prospective studies of oral contraceptives and breast cancer (Royal College of General Practitioners 1974; Vessey et al. 1975; Ory et al. 1976). These do not show an association, although follow-up periods were relatively short and the women subjects were relatively young. In view of the vast number of women who have taken oral contraceptives, longer term prospective studies are needed.

References

Arial IM (1973) Enovid therapy (norethynodrel with mestranol) for fibrocystic disease. Am J Obstet Gynecol 117:453–459

Boston Collaborative Drug Surveillance Programme (1973) Oral contraceptives and venous thromboembolic disease, surgically confirmed gallblader disease, and breast tumors. Lancet I: 1139–1404

Brown JM (1970) Histological modification of fibroadenoma of the breast associated with oral hormonal contraceptives. Med J Aust 1:276–277

Casey HW, Giles RC, Kwapien RP (1979) Mammary neoplasia in animals: pathologic aspects and the effects of contraceptive steroids. In: Lingeman C (ed) Carcinogenic steroids. Springer Berlin Heidelberg New York (Recent results in cancer research, vol 66, pp 129–160)

Dickey RP, Stone SC (1975) Drugs that affect the breast and lactation. Clin Obstet Gynecol 18:95–111

Drill VA, Martin DP, Hart ER, McConnell RG (1974) Effect of oral contraceptive on the mammary glands of rhesus monkeys: a preliminary report. J Natl Cancer Inst 52:1655–1657

Fechner RE (1970a) Fibroadenomas in patients receiving oral contraceptives: a clinical and pathologic study. Am J Clin Pathol 53:857–864

Fechner RE (1970b) Fibrocystic disease in women receiving oral contraceptive hormones. Cancer 25:1332–1339

Fechner RE (1977) Influence of oral contraceptives on breast diseases. Cancer 39: 2764–2771

Geil RG, Lamar JK (1977) FDA studies of estrogen, progestogens, and estrogen/progestogen combinations in the dog and monkey. J Toxicol Environ Health 3:179–193

Goldenberg VE, Wiggenstein L, Mottet NK (1968) Florid breast fibroadenomas in patients taking hormonal oral contraceptives. Am J Clin Pathol 49:52–59

Gould VE, Wolff M, Mottet NK (1972) Morphologic features of mammary carcinoma in women taking hormonal contraceptives. Am J Clin Pathol 57:139–143

Hamilton JM (1974) Comparative aspects of mammary tumors. Adv Cancer Res 19:1–37

Henderson BE, Powell D, Rosario I, Keys C, Hanisch R, Young M, Casagandi J, Gerkins K, Pike MC (1974) An epidemiologic study of breast cancer. J Natl Cancer Inst 53:609–614

Kelsey JL, Holford TR, White C, Mayer ES, Kilty SE, Acheson RM (1978) Oral contraceptives and breast disease: An epidemiological study. Am J Epidemiol 107: 236–244

Kirschstein RL, Rabson AS, Rusten GW (1972) Infiltrating duct carcinoma of the mammary gland of a rhesus monkey after administration of an oral contraceptive: A preliminary report. J Natl Cancer Inst 48:551–556

Kusama S, Spratt JS, Donegan WL, Watson FR, Cunningham C (1972) The gross rates of growth of human mammary carcinoma. Cancer 30:594–599

LiVolsi VA, Stadel BV, Kelsey JL, Holford TR, White C (1978) Fibrocystic breast disease in oral-contraceptive users. A histopathological evaluation of epithelial atypia. N Engl J Med 299:381–385

Nelson LW, Weikel JH Jr Reno FE (1973) Mammary nodules in dogs during four years treatment with megestrol acetate or chlormadione acetate. J Natl Cancer Inst 51:1303–1311

Nomura A, Comstock GW (1976) Benign breast tumor and estrogenic hormones: A population-based retrospective study. Am J Epidemiol 103:439–444

Norris HJ, Taylor HB (1970) Carcinoma of the breast in women less than 30 years old. Cancer 26:953–959

Ory H, Cole P, MacMahon B, Hoover R (1976) Oral contraceptives and reduced risk of benign breast diseases. N Engl J Med 294:419–422

Paffenbarger RS, Fasal E, Simmons ME, Kampert JB (1977) Cancer risk as related to use of oral contraceptives during fertile years. Cancer 39:1887–1891

Pearlman AW (1976) Breast cancer-influence of growth rate on prognosis and treatment evaluation. A study based on mastectomy scar recurrences. Cancer 38: 1826–1833

Penman HG (1970) The effect of oral contraceptives on the histology of carcinoma of the breast. J Pathol 101:66–68

Prechtel K, Seidel H (1973) Der Einfluß oraler Steroidkontrazeptiva auf das Fibroadenom der Mamma. Dtsch Med Wochenschr 98:698–702

Royal College of General Practitioners (1974) Oral contraceptives and health. Pitman, London

Royal College of General Practitioners' Oral Contraception Study (1977) Effect on hypertension and benign breast disease of progestogen component in combined oral contraceptives. Lancet I:624

Sartwell PE, Arthes FG, Tonascia JA (1973) Epidemiology of benign breast lesions: lack of association with oral contraceptive use. N Engl J Med 288:551–554

Sartwell PE, Arthes FG, Tonascia JA (1977) Exogenous hormones, reproductive history, and breast cancer. J Natl Cancer Inst 59:1589–1592

Thomas DB (1978) Role of exogenous female hormones in altering the risk of benign and malignant neoplasms in humans. Cancer Res 38:3991–4000

Vessey MP, Doll R, Sutton PM (1972) Oral contraceptives and breast neoplasia: A retrospective study. Br Med J 3:719–724

Vessey MP, Doll R, Jones K (1975) Oral contraceptives and breast cancer. Progress report of an epidemiological study. Lancet I:941–944

Wazeter FX, Geil RG, Cookson VR et al. (1976) Seven-year progress report on long-term oral contraceptive studies in female dogs and monkeys. Toxicol Appl Pharmacol 37:178

Pathomorphology of the Female Breast Induced by Exogenous Sex Hormones

R. Bässler and H. Schirmacher

Abstract

For the female breast to develop normally, several mammotropic hormones must act together during puberty to stimulate growth and differentiation of parenchymatous and mesenchymal components of the breast. When sex hormones act on a normally developed breast they induce parenchymal proliferation and secretion. Both represent the usual responses of the parenchyma to mammotropic hormones. The harmony of a physiologic stimulation, as in pregnancy or lactation, is generally lacking during accessory or exogenous hormonal action. As a result, definite patterns of tissue changes are brought about depending on the mechanisms of hormonal stimulation involved. Experimental studies show best the alterations induced by estrogens and gestagens. When administered continuously, they induce tubuloacinous development of the mammary glands with secretion, retention, calcification, fibrosis, and involution. Results of histometric studies are presented to demonstrate these changes. In comparison, it is much more difficult to evaluate the effects of accessory hormones on the human breast.

1. *Oral contraceptive agents,* according to our measurements, cause no enlargement of normal acini and no changes of the enlarged lobuli referred to as lobular hyperplasia. The diameter of lobules, the number of ductal sproutings and surface areas in women taking or not taking contraceptive agents are virtually alike.
2. *Galactorrhea* is frequently seen during use of contraceptive agents. Due to the action of several hormones, one sees histologically lobular hyperplasia with evidence of protein and fat synthesis, enlarged alveoli, dilatation of ducts, stasis of secretion and so-called swiss-cheese cysts. The galactostasis often originates a chronic (nonpuerperal) mastitis.
3. *Combined hormonal therapy* may induce lobular hyperplasia with secretion and the development of cysts. Although the patterns are irregular, they should not be regarded as those of fibrocystic disease (fibrocystic mastopathy). Under the stimulus of estrogen papillary proliferations appear in the lactiferous ducts, and prolactin induced the lobular secretion (formerly, residual lactation) with dilatation of the acini and development of lobular cysts. We encountered no cellular atypism of the ductal or lobular epithelial cells in our cases.

For the human female breast to develop normally, several mammotropic hormones must act in concert during puberty to stimulate the growth and differentiation of parenchyma and mesenchyma. The effect of accessory sex hormones on the normally developed breast, regardless of age, is to stimulate parenchymal proliferation, and often the secretion of a protein- and lipid-rich fluid. These responses are intimately connected and represent the parenchyma's most common reaction to mammotropic hormones. A physiological stimulation such as that in puberty, pregnancy, and lactation, induces morphologically a largely homogeneous tissue pattern. As a rule, the administration of exogenous hormones disturbs this nearly isomorphic picture, and we can expect to see ductal and lobular proliferation of a type depending on the mechanism of hormonal action. In childhood and before the onset of puberty these changes usually take the form of an enlargement of the entire organ, i.e., infantile hypertrophy or macromasty, whereas in the mature mammary gland changes in consistency, tumorlike indurations, or galactorrhoeal symptoms come to the fore.

It is difficult to chart the effects of an accessory hormone on the breast by pathohistological methods only, since numerical or histometric alterations of the lobules cannot be judged visually and since previous disorders (e.g., mastopathia cystica fibrosa, mastitis) may have changed the physiological pattern considerably.

However, experimental mamogenesis gives us a clear view of how and over what time period the gland may develop and involute, and provides an insight into some of the transformation processes that may prove applicable to the human case.

Experimental Mammogenesis

Developmental studies of the mammary gland have shown that neither estrogene nor progesterone by themselves cause a homogeneous differentiation of lobular parenchyme. Therefore combinations of these substances have been applied in the attempt to find the optimal physiological ratio for each species.

In castrated rats, estrogens have been found to induce continual growth of parenchyme up to the 10th day of administration; thereafter a low-grade involution ensued, despite continued injections, and went on to the 20th day. This involution could not be reversed even by further administration of estrogens for as long as 60 days. Stopping injections on the 20th day led to a rapid involution of the parenchyma and to a progredient and compensatory increase in fat tissue. Experiments with estrogen-progesterone mixtures (5 γ : 1 mg) have shown that they accelerate mammary development up to day 20; despite continued application, however, a slow regression followed. When injections were discontinued on day 20 the gland had regressed almost completely by day 40 (Blume 1970).

In histologic terms, we observe a sequence of proliferation of the ductal system with ductular sprouting and lobule formation; followed by secretion, dilation of the ducts, galactostasis, secret-retention with dyschylic alterations, calcification, and the formation of cysts accompanied by circumductal fibrosis. The histologic pattern mimics that of fibrocystic disease of the female breast.

Some reactions of human breast tissue to the administration of accessory mammotropic hormones are illustrated by the examples below.

Effect of Contraceptives on Breast-Parenchyma

Thanks to recent insights into the pathophysiology of contraceptives, we now have a fairly clear idea about some of their side effects. Yet we still have little information about how they affect mammary parenchyma (Erb and Kallenberger 1972), except perhaps with respect to certain disorders –: fibroadenoma, for instance, seems to occur less frequently in women who are taking contraceptives, which seem to have a stabilizing and prophylactic influence on the formation of this type of tumor (Prechtel 1969; Prechtel and Seidel 1972, 1973). Fechner (1970, 1972) has reported finding no histological differences between fibroadenomas in women taking and not taking contraceptives; on the other hand, Goldenberg et al. (1968) and Brown (1970) have related individual cases of serious epithelial proliferation to the action of these substances. In cases of fibrocystic mastopathy, Fechner (1972) again found no histologically detectable differences.

Investigations conducted by our own team were directed at the question whether women taking contraceptives showed alterations in normal lobules that we could detect histometrically or morphologically, even though they lay below the clinical threshold and did not lead to galactorrhea. In his study of mammary biopsies from 110 women aged 20 to 40, Burschel (1979) compared 46 cases of contraceptive use with the remaining women not on contraceptives. The women in the former group had taken the pill, predominantly one-phase substances, for 1 to 5 years before biopsies were taken for benign diseases. The histometric data obtained was compared to normal findings in women of the same age with benign diseases (Table 1). Burschel (1979) drew the following conclusions:

a) *Normal* lobules in sexually mature women have 25 to 34 sprouts, a mean diameter of 0.5 mm, and a surface area of 0.23 mm^2. Any lobules considerably larger than this were described as hyperplasia, i.e., when the number of sprouts had more than doubled and the surface area had increased by about half. Sprout number and surface area were proportional.

b) A comparison of the values for contraceptive users with non contraceptive users revealed *no* differences, either in normally differentiated lobules or in cases of lobular hyperplasia.

c) Contraceptives did not induce increased cellular proliferation in either normal or hyperplasic lobules. Lobular hyperplasias were found less frequently in women using contraceptives.

Table 1. Measurements of mammary lobules

Mean values	Without contraceptives		With contraceptives	
	Normal lobules	Lobular hyperplasia	Normal lobules	Lobular hyperplasia
Diameter (mm)	0.54	0.64	0.55	0.65
No. of sprouts	29.6 (25–34)	78.2 (70–89)	30.5 (23–37)	71.3 (55–97)
Surface area (mm^2)	0.23	0.32	0.24	0.33

d) Contraceptives bring no increased risk of benign breast diseases. In our sample we diagnosed fibrocystic disease (Mastopathia cystica fibrosa) more frequently in controls than in the women who had a history of contraceptive use.

Galactorrhea

Galactorrhea, defined as the spontaneous secretion of a milk-like fluid from one or both breasts at times other than pregnancy or puerperium, is often associated with amenorrhea and is a polyetiologic symptom.

The *pathogenesis of galactorrhea* has been largely explained over the past few years by investigations on prolactin metabolism by Wolstenholme and Knight (1972, lit.) and Von Werder (1975). Today we differentiate between two causes: a primary (pituitary, autonomous) hyperprolactinemia and a secondary (hypothalamic) form. The first type of hyperprolactinemia appears in conjunction with prolactin-forming pituitary tumors; the second, which has the same symptoms, is an expression of a disturbance in prolactin inhibiting factor (PIF) or an irritation of the hypothalamus. Since hyperprolactinemia leads to an inhibition of ovulation in fertile women, Von Werder (1975) has suggested the term "galactorrhoea-amenorrhea syndrome." Here, hyperprolactinemia triggers galactorrhoea when mammary epithelium is stimulated simultaneously by estrogens and gestagens under physiological conditions. A rise in prolactin level coupled with steroid concentration leads in prolactin-sensitive organs, of which the mammary gland is one, to an increase in prolactin receptors (Von Werder and Rjosk 1979). In 70% of their patients with hyperprolactinemia the authors found galactorrhoea, which as a rule was caused by pressure on the perimammillary tissue. Spontaneous galactorrhoea is much rarer. But even at absolutely normal prolactin levels, galactorrhoea may appear, which indicates that the tissue has become more sensitive to prolactin in physiological concentrations. Wyss et al. (1971) found galactorrhoea as an isolated symptom in 0.6%, and in conjunction with amenorrhea in 0.12% of cases. Pathogenesis and treatment of the various galactorrhoea syndromes have been described by Bässler (1978) and Von Werder and Rjosk (1979).

There are three main types of *galactorrhoea induced by exogenous hormones:*

Contraceptive-Induced Galactorrhoea

Oral contraceptives sometimes trigger cyclic, if low-grade galactorrhoea, which may be traced to a lowering of FSH and LH levels in the blood and a decrease in "prolactin inhibiting factor" (PIF) accompanied by temporary hyperprolactinemia. The first observations were made by Gregg (1966) and Rosen et al. (1967). Since then a number of other studies have been published (Tapia and Haefeli 1978).

In pathohistological views, the smears of mammary secretion showed homogeneous and cell-poor protein and lipid precipitates that contained foam cells. The parenchyma is characterized by focal enlarged lobules with disaggregated intralobular connective tissue and enlarged acini, some of which had been transformed into alveoli as a result of secretion.. These reactions were considerably weaker than during lactation, and corresponded approximately to the mammary's state during the second

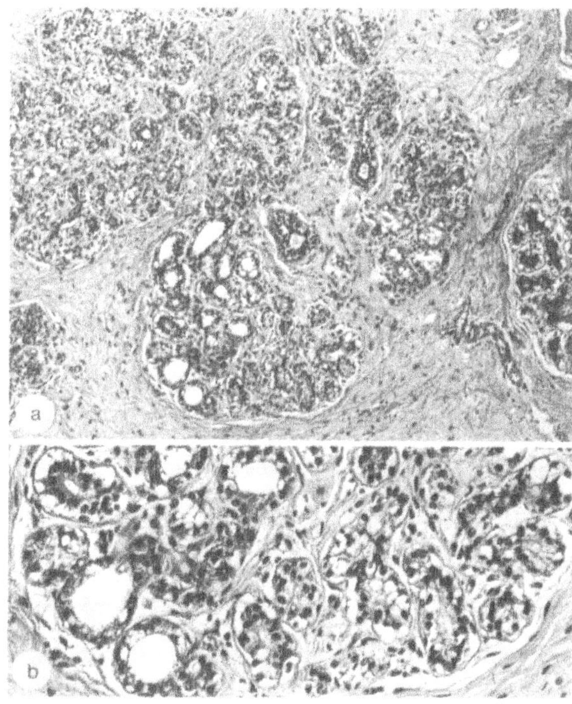

Fig. 1. Lobular hyperplasia showing signs of secretion formation in lobular cells (light-colored lipid vacuoles) and secretion. Note alveoli formation. From a patient with galactorrhea and a long history of contraceptive use. **a** Overview, **b** detail of some lobules. Form., Paraff., HE, × 45.5 and × 149.5

trimester of pregnancy, though in this case the reactions were not homogeneous. Galactorrhoea here can be traced to this transformation of the lobules, as can be seen from Fig. 1.

Case reports: One of our own patients, a 30-year-old woman who had been taking contraceptives since 1966 (Ovulen, then Mikrogynon and Edival to 1975), had galactorrhea for the past 2 years. A biopsy of the left breast for a suspected tumor with diffuse boundary was performed; microscopically it showed lobular parenchyma in various stages of differentiation that contained areas with secretion (Fig. 1). Tumor was not confirmed histologically. Induration was due to lobular hyperplasia with secretion and secret-retention in the lactiferous ducts. There were no signs of infection.

Figure 2 illustrates a nodular adenofibroma of the mammary, showing discrete secretion but no galactorrhea. This specimen is from a 42-year-old woman who had been taking contraceptives continually for 13 years when she noticed a tumor in her left breast. Histologically, this was a glandular and tubular adenofibroma with partial lobular division and a sharp pseudocapsular boundary. It showed, however, a low rate of secretion and retention, very probably the result of years of contraceptive use. The tubular proliferations also point to estrogen stimulation (Fig. 2).

Drug-Induced Galactorrhoea

This type has gained in importance in the past years, since we now know that galactorrhea may be a sideeffect of psychopharmaca, especially phenothiazins (reserpin and methyldopa). These substances suppress the prolactin inhibiting factor (PIF) and in-

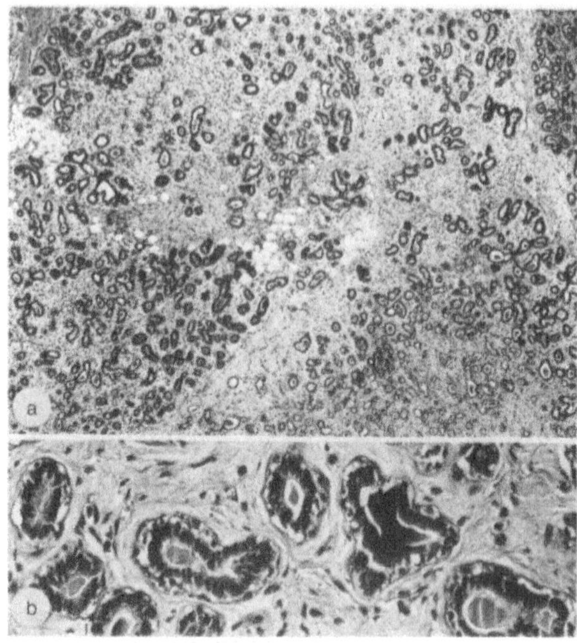

Fig. 2. a Lobular adenofibroma with discrete secretion, showing large numbers of crowded, narrow ducts with sprouts in transparent connective tissue. From a patient who had taken contraceptives for 13 years. **b** Detail of same. Form., Paraff., HE, x 45.5 and x 149.5

duce galactorrhoea the extent of which depends on dose and length of medication (Bässler 1978). In an experimental setting, these substances in combination with estrogens can be shown to induce lactation that, seen under the electron microscope, is characterized by a secretory transformation of the mammary-gland-cells (Pier et al. 1970).

Case reports: One of our patients, a 39-year-old woman who had her menarch at 13, regular menstruations of 28/5 days, and no pregnancies, had taken contraceptives from her 30th to 32nd year. She had been treated for schizophrenia for over 10 years with daily 50 to 100 mg of Limbatril (amitriptylin and chlordiazepoxid) and Taraktan (chlorprothixen). She presented with nodular induration of the left breast and discrete secretion, therefore she unterwent a clinical examination and mammography. The findings were a suspected carcinoma with microcalcifications and cysts. Subcutaneous mastectomy was performed. The specimen measured 8 x 9 x 2.5 cm, had a roundish shape and varying consistency. A walnut-sized cyst filled with fluid projected from its surface. Incisions revealed multiple small cysts, increased tissue density and strong secretion in some areas.

We observed a proliferative fibrocystic disease with small and large cysts but no cellular dysplasias or atypism. In addition to normal lobules and cystic alterations we noted many areas of enlarged lobules with alveolar transformation due to secretion and retention. Some of the lobule groups were unchanged (Fig. 3). The enlarged alveoli contained a nonhomogeneous, protein- and lipid-rich secret with masses of desquamatous cells and foam cells (Fig. 3b, c). Also present were smaller alveoli with button-like, projecting glandular cells (Fig. 3b) ("cells resembling tennis rackets with apocrine nuclear projection", Hamperl 1975). There were also signs of merocrine secretion. These

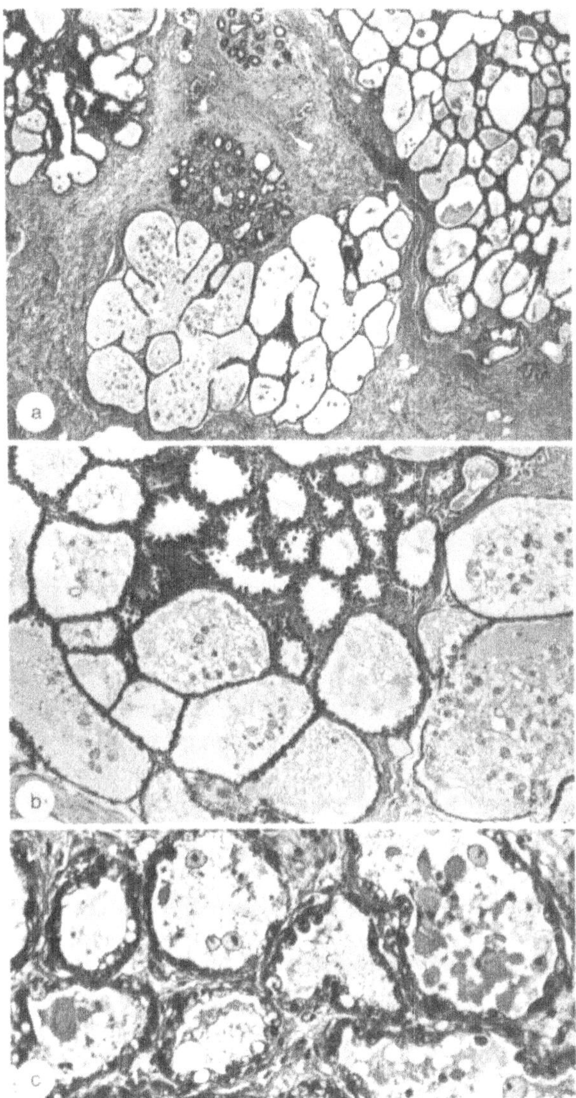

Fig. 3. a-c. Effect of accessorius prolactin on mammary tissue, showing inhomogenous secretion in lobules, alveolar transformation, and galactostasis. **a** Overview showing also some unaltered lobules. **b** Detail showing uneven secretion and so-called lobular secretion in the center. **c** Detail showing secret formation, retention, and foam cells. Form., Paraff., HE, x 45.5, 149.5, 182

regions correspond to the "lobular secretion" seen in some cases of fibrocystic disease, and which McFarland (1922) called "residual lactation."

In addition to the effects of fibrocystic disease, the breast evinced unusual alterations in numerous lobules, secret-retention, and single lobules button-shaped protrusions of the epithelial cells. These changes resulted from prolactin stimulation accompanying galactorrhoea, which in turn arose from suppression of prolactin inhibiting factor as a side-effect of treatment using psychopharmaca. Prolactin level at the time of examination was 62 ng/ml (serum), about that seen during the second trimester of pregnancy.

In another case, of a 56-year-old woman, with metastasizing breast cancer who had been on estrogen-gestagen therapy, post mortem examination showed strong secretion in the contralateral mammary, galactorrhoea, and swiss-cheese cysts (Bässler 1978). We also noted that the accessory sex hormones had stimulated the contralateral breast.

Similar changes have been seen in *male mammary glands* after hormone therapy for prostata carcinoma. Administration of estrogens has led to lobular gynecomasty with secretion, and galactosis with swiss-cheese cysts. This may explain the cases of "lactating men" described in the early literature (Bässler 1978).

Galactorrhoea as a Facultative Symptom in Hormone Therapy

Since this paper focuses on pathohistologic changes in breast parenchyma effected by exogenous hormones, we shall treat the finding of galactorrhoea separately in the discussion below. This category is taken to include the lobular gynecomastia with secretion often seen in men when they receive hormone therapy for prostata cancer.

Mammary Pathohistology as an Incidental Effect of Hormone Therapy

Knowing the normal biomorphosis of the mammary gland and the differences between its various sectors, we often found tissue patterns in our biopsy specimens that diverged from the rule. Among these were lobular hyperplasia, symptoms of secretion, secretion retention with ductal ectasia, galactostatic cysts, and a chemically induced mastitis when the secret had penetrated the circumductal mesenchyma through ruptures in the duct walls. These alterations correspond to what is known as "retention syndrome" in older women with ductal ectasia and "plasma cell mastitis." The observations discussed here were made generally in younger women whose parenchyma showed the effects of estrogen and gestagen treatment, particularly secretory transformation of the enlarged lobules and the results of a missing or insufficient drainage of this secretion. Therefore, galactorrhoea cannot be said to be a ruling symptom, but may be provoked when a slight pressure is applied on the breast, especially on the circumareolar region.

Case reports: A 37-year-old woman who had last given birth 13 years ago and had been suffering from hypermenorrhea for 3 years, had been on hormone therapy and contraceptives for a year. A few weeks before the biopsy was scheduled her left breast began to swell painfully, and she developed hyperemia of the skin and low-degree secretion. A biopsy was performed, and a cloudy, yellowish-grey fluid emerged. Pathohistologically, the diagnosis was purulent mastitis with secretion; also lobular hyperplasia as shown by low-degree secretion in lobules, partial retention and ectatic, small lactiferous ducts. Also present were intraductal pseudopapillary epithelial proliferation resulting from several years' hormone treatment, particularly with estrogens (Fig. 4).

Another patient, a 26-year-old girl, had been treated for secondary amenorrhea with progesterone (Primolut) and Progylut (Ethinylestradiol). A fibroadenoma was removed from her left breast; it showed no histological particularities. Thereafter nodules reformed, but these were less an indication of fibroadenoma than of fibrosis shot

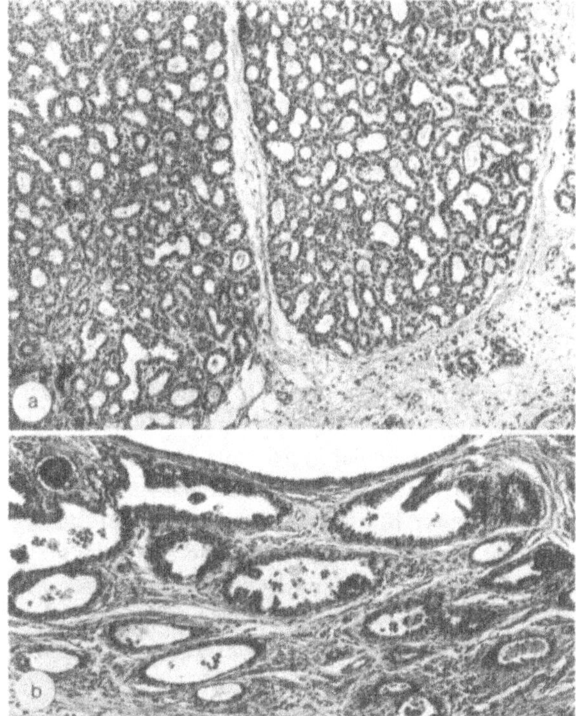

Fig. 4. a Lobular hyperplasia and discrete secretion in a case with purulent, nonpuerperal mastitis, after accessory hormone therapy. **b** Small lactiferous ducts showing intraductal, pseudopapillary epithelial proliferation. Form. paraff., HE, x 45.5 and x 136.5

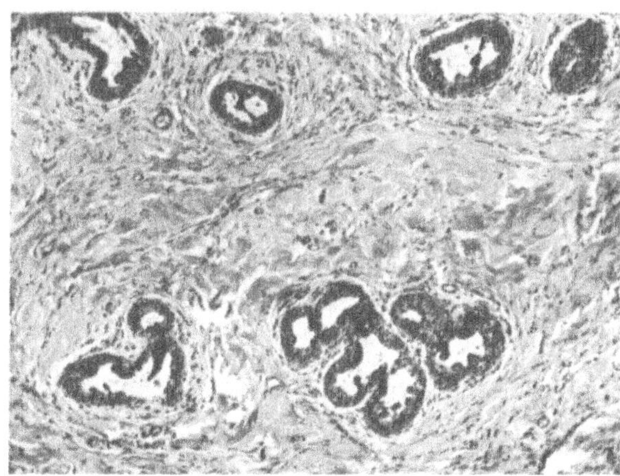

Fig. 5. Group of small lactiferous ducts with intraductal, often pseudopapillary epithelial proliferation, surrounded by dense stroma (example of typical gynecomasty). Form., Paraff., HE, x 148.4

through with tubular ductules which showed pseudopapillary epithelial proliferations. These changes look like the gynaecomastia, and could be symptom of accessory hormone effect (Fig. 5).

In contrast to the almost homogeneous proliferation of parenchyma seen during pregnancy and lactation, the breast usually reacts to exogenous hormone in a nonuniform manner, i.e., not all of the lobules and lobular segments show this transormation to an equal degree. On the other hand, we have observed cases of extremely heavy lobular secretion similar to that seen in fibrocystic disease (Bässler 1978), though in contrast to that condition great numbers of lobules were affected. This probably indicates that special prolactin receptors are present here which are activated more and more as steroid concentration from exogenous sources increases (Fig. 3a).

In making our *histological diagnosis*, we should keep the following two factors in mind:

1. Secretory transformation with galactostasis may mimic or superimpose itself on a *fibrocystic* pattern, and it would seem important to differentiate between the two. In fibrocystic disease actively secreting lobules are present in only low numbers, i.e., lobular secretion is limited, and the same is true of lipid synthesis in epithelial cells. When secretion is found in large numbers of lobules, as it is in galactostasis and mastitis, perhaps this is due to accessory hormone effects.
2. Hormone-induced hyperplasias and secret retention can cause *chronic abacterial mastitis* if the secret penetrates the duct wall. Thus when diagnosing the various types of chronic mastitis one should check whether a lobular hyperplasia with signs of secretion retention is present. There is no question that galactostasis and enlargement of the lactiferous ducts can lead to bacterial infection; however, in cases with galactostasis and duct ectasia (retention syndrom) we observe usually chronic and abacterial mastitis. Mühlenstedt and Schneider (1976) have reported on their clinical experience with this condition, and Peters and Breckwoldt (1978) on three further observations of galactorrhea with mastitis. According to these authors, improvement was brought about without having to use antibiotics, solely by administering Bromocriptin (see also Fig. 6).

Figure 6 gives a schematic view of the pathogenetic relationships between galactorrhea, galactostasis and mastitis.

Fig. 6. Pathogenetic relationships between hormone-induced galactorrhea, galactostasis and nonbacterial mastitis

Summary

Mammotropic hormones that have an accessory effect on the normal mammary gland cause, independently of age and sex, to a proliferation of the parenchyma and often to the secretion of a protein-containing and lipid-rich liquid. These reactions, which are intimately connected, are the cell's normal answer to hormonal stimulation of its receptor mechanisms. Depending on dose and duration in application, accessory hormones bring about inhomogeneous proliferation patterns that, in childhood, take the form of infantile hypertrophy (macromasty) and in adulthood often produce changes in tissue consistency or galactorrhoea. Experimental mammogenesis shows these estrogen and gestagen effects in other species.

The pathohistology of incidental hormone effects was investigated in 15 cases with the following results:

1. In women 20 to 40 years of age, contraceptives effected no enlargement of lobules and no lobular hyperplasia in comparison to women not on medication.
2. Galactorrhea, a polyetiological symptom, was observed after contraceptive use, after administration of certain drugs (phenothiazines), and as a symptom accompanying therapy with exogenous sex hormones.
3. The administration of exogenous mammotropic hormones leads, as a rule, to inhomogenous tissue patterns characterized by ductal and lobular hyperplasia, secret retention syndrome with galactostatic cysts, and chronic mastitis. When making differential-diagnosis of unusual and secretory active forms of fibrocystic disease, chronic mastitis or lobular gynecomastia, one should consider the possibility that these may arise as hormone sideeffects.

References

Bässler R (1978) Pathologie der Brustdrüse. In: Doerr W, Seifert G, Uehlinger E (eds) Spezielle pathologische Anatomie, vol 11. Springer Berlin Heidelberg New York, pp 1–1134

Blume G (1970) Morphologische und histometrische Untersuchungen an der Brustdrüse der Ratte bei hormonaler Stimulation und Involution. Inaugural Dissertation, Universität Mainz

Brown IM (1970) Histological modification of fibroadenoma of the breast associated with oral hormonal contraceptives. Med J Aust 1:276–277

Burschel C (1979) Das lobuläre Parenchym der weiblichen Brustdrüse unter physiologischen Bedingungen und bei Einnahme von Kontrazeptiva. Ergebnisse histometrischer Untersuchungen. Inaugural Dissertation, Universität Marburg

Erb H, Kallenberger A (1972) The action of an high-dosed oestrogen-progestagen-combination on the human breast. Acta Endocrinol (Kbh) 70:143–155

Fechner RE (1970) Fibroadenoma in patients receiving oral contraceptivas: a clinical and pathologic study. Am J Clin Pathol 53:857–864

Fechner RE (1972) Benign breast disease in women on estrogen therapy. A pathologic study. Cancer 29:273–279

Goldenberg VE, Wiegenstein L, Mottet NK (1968) Florid breast fibroadenoma in patients taking hormonal oral contraceptives. Am J Clin Pathol 49:52–59

Gregg WI (1966) Galactorrhoea after contraceptive hormons. N Engl J Med 274: 1432–1433

Hamperl H (1975) Sekretionserscheinungen in der mastopathischen Brustdrüse. Virchows Arch. Abt. B 18:73–81

McFarland J (1922) Residual lactation acini in the female breast. Their relation to chronic cystic mastitis and malignant disease. Arch. Surg. 5:1–64

Mühlenstedt D, Schneider HPG (1976) Chronisch-rezidivierende Mastitis bei Galaktorrhoe. Geburtshilfe Frauenheilkd 36:1102

Peters F, Breckwoldt M (1978) Nonpuerperale Mastitis bei hormonal-stimulierter Mamma. Kasuistischer Beitrag. Geburtshilfe Frauenheilkd 38:754–757

Pier WJ, Garancis JC, Kuzma JF (1970) Fine structure of tranquilizer-induced changes in vat mammary gland. Am J Pathol 60:119–130

Prechtel K (1969) Ovulationshemmer und Brustdrüsenveränderungen bei Frauen im geschlechtsreifen Alter. Münch Med Wochenschr 111:2443–2447

Prechtel K, Seidel H (1972) Brustdrüsenveränderungen nach Langzeitbehandlung mit sogenannten Ovulationshemmern. Verh Dtsch Ges Pathol 56:529–532

Prechtel K, Seidel H (1973) Der Einfluß oraler Steroidkontrazeptiva auf das Fibroadenom der Mamma. Dtsch Med Wochenschr 98:698–702

Rosen SW, Gahres EE (1967) Nonpuerperal glactorrhoea and the contraceptive pill. Obstet Gynecol 29:730–731

Tapia JE, Haefeli MM (1978) Verändern hormonale Kontrazeptiva die weibliche Brust? DIA Mod Ther 2:102–109

Werder K von (1975) Wachstumshormone und Prolaktinsekretion des Menschen. Physiologie und Pathophysiologie. Urban, München Berlin Wien

Werder K von, Rjosk HK (1979) Menschliches Prolaktin. Klin Wochenschr 57:1–12

Wolstenholme GEW, Knight J (eds) (1972) Lactogenic hormones, A CIBA Foundation symposion in memory of Prof. S.J. Folley. Churchill Livingstone, Edinburgh London

Wyss HJ, Krauer F, Vetter L (1971) Die Galaktorrhöe. Schweiz. Z. Gynäk. Geburtsh. 2:117–124

Estrogen Receptors and Ultrastructural Pathology of Mammary Carcinoma

H.-E. Stegner, H. Maass, G. Trams, and C. Pape

Abstract

The interrelationships of structural and biochemical parameters have been analysed in a study of 342 cases of invasive breast carcinoma. There was no differential relationship between the histologic subclassification and distribution of estrogen receptors. However, significant correlations could be found between estrogen binding capacity and ultrastructural features of the tumor cells determined on 118 specimens. Estrogen receptors were more frequently associated to tumors of high grades of cellular differentiation showing great complexity of cytoplasmic organization, high degrees of cell interconnections by desmosomes and membrane interdigitations, presence of specific secretory granules and low grades of nuclear pleomorphy. Obviously the development of estrogen receptors goes parallel to the individual grade of cell differentiation and is independent from ductal or lobular histogenesis.

Introduction

The finding of steroid receptors on normal and neoplastic cells has led to a better understanding of tumor growth and given us a more precise basis on which to treat steroid-sensitive carcinomas with hormones. Steroid receptors can be found in 50% to 70% of primary mammary cancers; the rate for metastatic tumors is lower. In approximately 60% of cases where steroid receptors are present, endocrine therapy can bring an objective remission; whereas in cases without receptors the probability of remission is less than 10% (Maass et al. 1975; McGuire et al. 1975; Heuson et al. 1977).

When we categorize breast cancers as receptor-positive or receptor-negative, we naturally begin to ask whether there might be some connection between receptor presence and the histologic type of tumor. However, we run into difficulties with almost all the common schemes for subclassifing these tumors, because they mix criteria – topographic, histogenetic, gross morphological, fine structural, and prognostic. This is apparently the reason why no consistent correlation has yet been found between receptor presence and histologic tumor type (Terenius et al. 1974; McGuire et al. 1975; Rosen et al. 1975, 1977; Wittliff et al. 1976).

By means of an ongoing, prospective study we have attempted to find a correlation between ultrastructural organization and receptor presence in mammary carcinoma[1]. From a total of 342 declared cases we have looked at 118 tumors under the elec-

---

1 Study supported by DFG. Cooperative Prospective Studies of Mammary Carcinoma Ma 310/9 and Ste 91/8

Table 1. Incidence rate of histological tumor types

Solid	120	(34.3%)
Glandular	148	(42.3%)
Infiltrating lobular	30	(8.6%)
Medullary	15	(4.3%)
Colloid	12	(3.4%)
Cribriform	4	(1.1%)
Intraductal comedo	2	(0.6%)
Squamous cell	2	(0.6%)
Papillary	1	(0.3%)
Cystosarcoma (malignant)	1	(0.3%)
Unclassified	7	(2.0%)

Table 2. Ultrastructural Parameters of Differentiation

Organelle content	Cell binding	Nuclear structure
Mitochondria	Desmosomes	Polymorphy
Ergastoplasma	Semi-desmosomes	Heterochromatin
Ribosomes	Thight junctions	Nuclear inclusions
Golgi structures	Interdigitation	Macronucleoli
Filaments	Basal membrane	
Secretory granules	Intracytoplasmic ductules	
Glycogen		
Lipids, lysosomes		

tron microscope. Rather than characterizing them according to histologic type (Table 1), we defined them semiquantitatively according to the parameters of Table 2 and entered the results on a survey sheet. Then we performed a statistical analysis to correlate ultrastructural characteristics with biochemical findings, and checked the mean differences for significance using the *t*-test, and frequency deviations with the aid of a chi-square test. Correlations of the order of 0.05 were considered to be statistically significant.

Estrogen receptor concentration in freshly prepared samples was determined by measuring the ^{3}H–17β-estradiol binding capacity in the cytosol fraction. The steroid receptor complex was analyzed using agar gel electrophoresis. Binding capacity is given in fmol/mg of tissue protein.

Fine Structure of the Ductoalveolar System

First, a short rundown on the cytomorphology of the mammary cell might throw our pathologic findings into sharper relief. The epithelium of the lactiferous ducts and lobules consists of two layers, a surface layer of cuboid or prismatic cells and a discontinuous layer of basal cells which is bordered by the stroma of the basement membrane

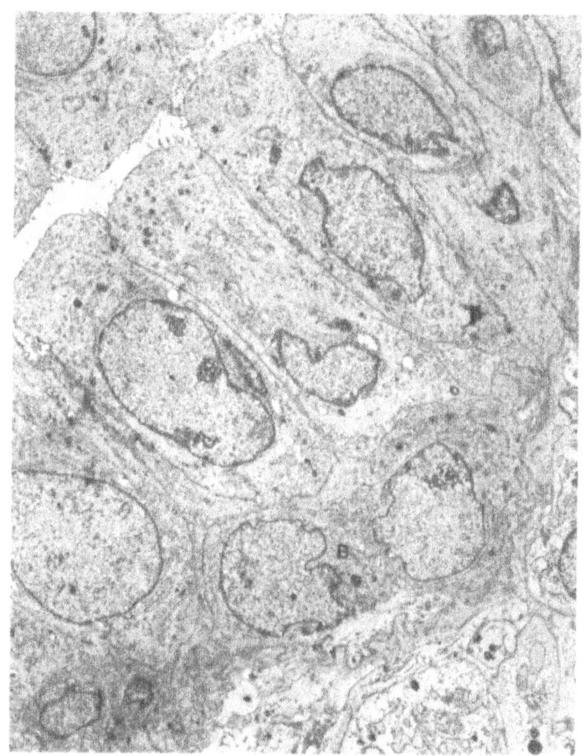

Fig. 1. Normal duct, showing two-layered epithelium. *B* = Basal cells. x 3,372

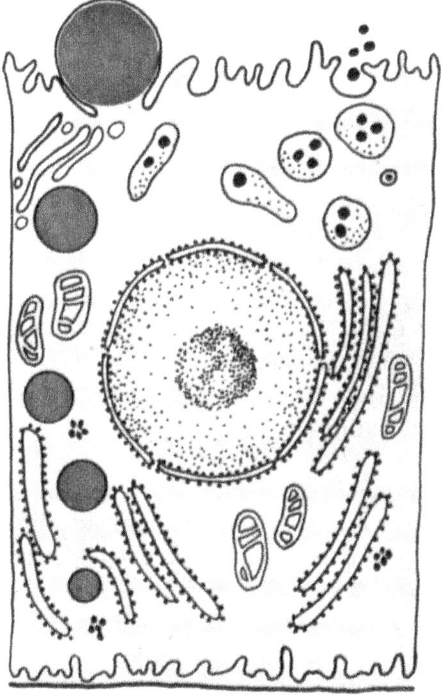

Fig. 2. Mammary gland structures involved in the synthesis of milk fats and protein

(Fig. 1). The surface cells have a light-colored cytoplasm and a round or oval nucleus. They are connected to neighbouring cells by desmosomes. The apical cell surfaces carry microvilli. Their cytoplasm contains mitochondria, free ribosomes, ergastoplasma, and Golgi structures. Isolated droplet of lipid and filaments are also found. Surface cells may be divided into two groups by size and organelle content: Light, organelle-poor main or B cells, and dark, organelle-rich A cells.

The basal cell layer is called the myoepithelium. It consists of cuboid or flat cells that lie along the basement membrane and contain dense bundles of filaments. Since they have a diameter of 50 to 80 Å and a Z-membrane type structure, these fibers resemble actin filaments. The basal, myoepithelially differentiated cell layer is the germination layer for the replacement of glandular epithelium. In their undifferentiated state basal cells contain only a minimum of organelles (Ozello calls them indeterminate cells).

The fine structure of alveolar cell is subject to cyclic changes; during galactogenesis and lactation they reach their peak of specialization. In order to form milk proteins and lipids they require special structures, particularly ribosomes, ergastoplasma, and Golgi vesicles (Fig. 2). Milk proteins are formed from amino acids from blood plasma. The activated amino acids flowing into the cell are transported via messenger RNA to the ribosomes and ergastoplasma, in whose structures preproteins are synthesized and transported towards the cell apex. In the Golgi apparatus, casein micelles agglomerate into definitive secretion granules. After being enclosed in cytoplasmic vacuoles they are ejected by exopinocytosis. The milk lipid is prepared within vesicles of smooth endoplasmic reticulum. These membranes merge upon secretion with the cell membrane, preventing confluence, and the granules are secreted as membrane-bound droplets.

Fine Structural Variations in Mammary Carcinoma

Depending on their degree of differentiation, breast cancer cells contain more or less complete sets of the typical organelles. One often sees combinations of organelles that are functionally connected to perform a certain type of synthesis and which reflect the cells' abortive specialization. As in normal epithelium, we find in many carcinomas a mixed population of organelle-poor (light) and organelle-rich (dark) cells (Fig. 3). These combinations are found in both lobular and ductal types. In lobular tumors, which are more often de-differentiated and immature than ductal tumors, the cells often contain only the most basic vital organelles. Here the monoform tumor cells resemble the undifferentiated basal cells of normal epithelium.

By contrast to these anaplastic carcinomas we have those that are characterized by a rich selection of organelles and a high degree of intracellular organization. Not infrequently their nuclei may even show a relatively homogeneous distribution of chromatin. In such case we usually classify them ultrastructurally as neoplastic by looking more at tissue relationships than at cytomorphological criteria. In principle, all the various specific structures of the mammary cell can appear in all common types of mammary cancer. Only their incidence seems to be type-specific. For example, we find *filaments* more frequently in lobular than in ductal carcinomas. *Secretion granules*

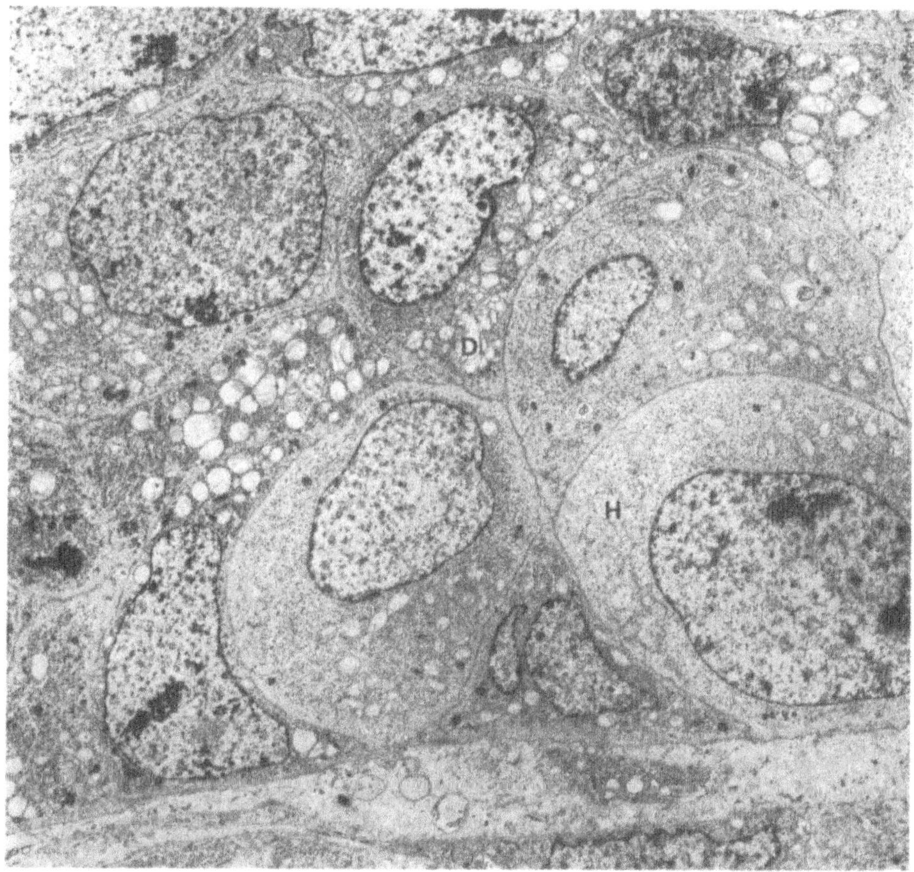

Fig. 3. Medullary breast cancer. *H* = light-colored, cytoplasm-rich cells. *D* = dark-colored cells. x 3,950

similar to casein particles, by contrast, appear more often, usually at the cell apex, in highly differentiated ductal tumors; and mucus vesicles are typical in colloid cancer. *Glycogen* is found more often in ductal than lobular types. The largest glycogen deposits we have yet observed were in a squamous cell cancer (Fig. 4).

Strong cellular binding is an indication of good coherence and a low dispersion tendency in cancer cells. *Membrane interdigitation* and relatively numerous *desmosomes* are found primarily in the group of breast cancers that show special differentiation. In these we have observed all forms of epithelial cell contact, from dense mosaics to intense interdigitation all the way to strong binding by means of specialized areas of cell membranes (zonulae occludentes, desmosomes).

The cells of invasive carcinoma may be dispersed within the stroma, or form solid and glandular complexes surrounded by a *basement-membrane*, detectable under the electron microscope. What this means is that the basement-membrane is no barrier to invasion, but a secretion product that borders epithelial and mesenchymal cells. A con-

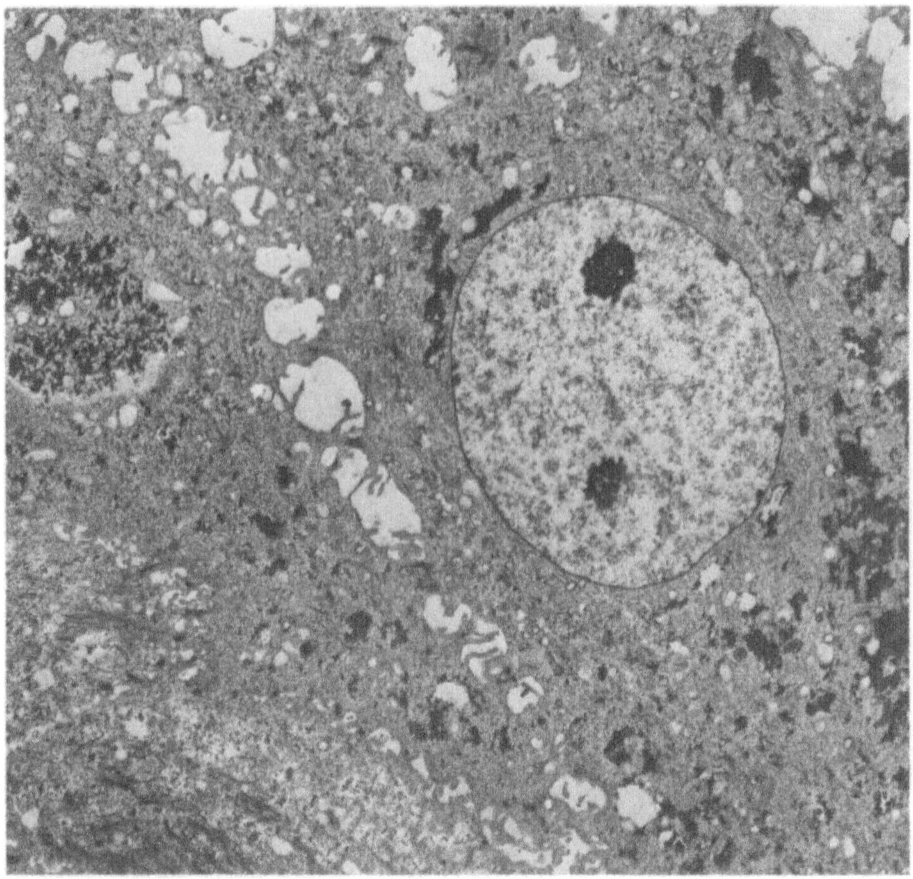

Fig. 4. Squamous cell carcinoms of the breast, showing large glycogen deposits in the tumor cells. × 4,589

tinuous basement membrane is more frequently found in glandular than in solid tumor types.

The meaning of the *intracytoplasmic ductules* is unknown (Fig. 5). They are found more frequently in lobular carcinomas, and indicate a high membrane activity that increases metabolism by dint of an enlargement of cell surface area. As regards *nuclear polymorphism*, we see a broad spectrum ranging from round and oval-shaped nuclei that contain finely granular and homogeneously distributed chromatin, to bizarre shapes with large, dense heteropyknotic areas. Sometimes the nuclei contain membrane-bound lamellar inclusions (caryosomes). The meaning of these structures is not known. The *nucleoli* are usually typically reticular in construction; condensation of nucleoli may indicate degeneration. Multiple nucleoli and macronucleoli are sometimes observed.

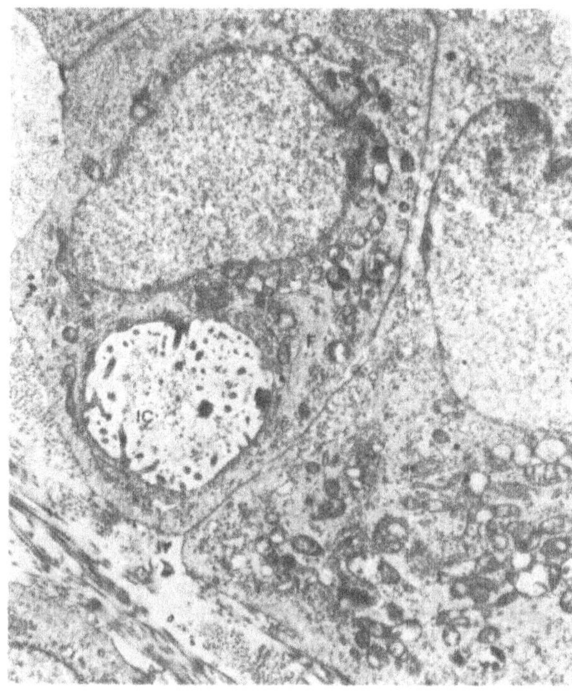

Fig. 5. Infiltrating lobular carcinoma of the breast. *IC* = intracytoplasmic canaliculus; *F* = filaments. x 8,010

Biochemical-Ultrastructural Correlations

We correlated the above-mentioned ultrastructural characteristics of 118 tumors with the results of our biochemical investigation for steroid receptors. The most important correlations may be summed up as follows.

Estrogen binding capacity of tumor tissue as determined by gel electrophoresis showed no significant correlation to histologic tumor type. However, the solid and highly fibrous types, i.e., those that contained high percentages of connective tissue, were less frequently receptor-positive than the other types. The group of cancers with unusual differentiation, to which as a rule the more highly differentiated types belong, registered receptor-positive more frequently. A higher proportion of ductal than lobular cancers were receptor-positive, though the difference was not significant.

Our findings agree with a trend that has been reported in a number of biochemical and histopathologic studies, e.g., Rosen et al. (1975, 1977) and Wittliff et al. (1976). In Wittliff's series, 50% of infiltrating carcinomas (not otherwise specified) had steroid receptors, and he found a similar percentage in his group of infiltrating lobular cancers. A subclassification of infiltrating ductal carcinomas revealed a concentration of receptor-positive tumors in the group of tubular and papillary cancers. By contrast, only 2 of 11 medullary were receptor-positive.

Our study revealed statistically significant correlations of receptor content with the ultrastructural parameters listed (total 118):

Marked nuclear polymorphy, more frequent in receptor-negative tumors ($P < 5\%$).
Heterochromatization, more frequent in receptor-negative tumors ($P < 5\%$).
Accompanying lymphocellular infiltration, more frequent in receptor-negative tumors ($P < 5\%$).
Mitoses, more frequent in receptor-negative tumors ($P < 1\%$).
Cells rich in mitochondria and ribosomes more frequent in receptor-negative tumors ($P < 5\%$).
Membrane interdigitation, more frequent in receptor-positive tumors ($P < 5\%$).
Golgi structures, more frequent in receptor-positive tumors ($P < 5\%$).
Synthesis products (secretion granules, lipids, mucus vesicles), more frequent in receptor-positive tumors ($P < 5\%$).
Intracytoplasmic ductules, more frequent in receptor-positive tumors ($P < 1\%$).

The following conclusions may be drawn from our findings: Tumors which have a high degree of ultrastructural differentiation (e.g., high intracellular organization, good inter-cell binding, specific synthesis products) are more frequently receptor-positive than undifferentiated tumors. The formation of steroid receptors apparently runs parallel to degree of maturation (differentiation) and is independent of histologic type as commonly classified. The histogenesis of the tumor, and particularly whether it has derived from the ductal or lobular compartment of the mammary gland, seems to have no influence on steroid receptor formation.

References

Heuson JC, Longeval E, Mattheiem WH, Deboel MC, Sylvester RJ, Leclercq G (1977) Significance of quantitative assessement of estrogen receptors for endocrine therapy in advanced breast cancer. Cancer 39: 1971

Maass H, Trams G, Nowakowski H, Stolzenbach G (1975) Steroid hormone receptors in human breast cancer and the clinical significance. J Steroid Biochem 6: 743

McGuire WL, Carbone PP, Vollmer EP (1975) Estrogen receptors in human breast cancer. Raven, New York

Ozello L (1971) Ultrastructure of the human mammary gland. Pathol Annu 6: 1

Rosen PP, Menendez-Botet CJ, Nisselbaum JS, Urban JA, Mike V, Fracchia A, Schwartz MK (1975) Pathological review of breast lesions analyzed for estrogen receptor protein. Cancer Res 35:3187

Rosen PP, Menendez-Botet CJ, Urban JA, Fracchia A, Schwartz MK (1977) Estrogen receptor protein (ERP) in multiple tumors specimens from individual patients with breast cancer. Cancer 39: 2194

Terenius L, Johansson H, Rimsten A, Thosen L (1974) Malignant and benign human mammary disease: estrogen binding in relation to clinical data. Cancer 33: 1364

Wittliff JL (1974) Specific receptors of steroid hormones in breast cancer. Semin Oncol 1: 109

Wittliff JL, Beatty BW, Savlov ED, Patterson WB, Cooper RA (1976) Estrogen receptors and hormone dependency in human breast cancer. In: Arneault GS, Band P, Israel L (eds) Breast cancer: A multidisciplinary approach. Springer, Berlin Heidelberg New York, p 59–77

Lobular Pattern of Mammary Tissue After Long-term Use of Combined Hormonal Contraceptives

K. Prechtel, U. Pöschl and K. Laakso

Abstract

By means of a prospective, morphological study covering several years we have been able to show that the relative risk of fibroadenoma, fibrocystic disease and carcinoma is not increased by long-term oral contraceptive use. There likewise was no increase in the rate of recurrence of fibroadenoma or fibrocystic disease. Qualitative and quantitative histomorphological analysis indicated that though long-term use of contraceptives led to a generally higher differentiation of the lobular system, it did not lead to secretory activity of the kind seen in mamma lactans. However, the changes oberseved were too varied and discrete to allow us to infer contraceptive use from tissue pattern type. Mammary gland tissue is thus not as acutely sensitive to steroids as endometrium.

Over a period of several years we have conducted a prospective, morphological study to determine whether the long-term use of hormonal contraceptives might lead to iatrogenic breast disease, in particular to fibroadenoma, fibrocystic disease, and carcinoma.

Based on biopsies, we have attempted to find the cause of these disorders as well as determine the relative risk of contracting them (Hartmann 1978). We compared a

Table 1. Relative risk of fibroadenoma, fibrocystic disease and carcinoma of the mammary gland[a]

Diagnosis	Pill-takers No.	%	Risk[b]
Fibroadenoma	(67/264)	25.4%	1.00
Fibrocystic disease	(153/987)	15.5%	0.61
Carcinoma	(22/275)	8.0%	0.31
Control	$\frac{33138}{129946}$	25.5%	1.00

[a] In women up to 55 years of age under long-term contraceptive use (period of observation 1972–1976)

[b] $\frac{\text{\% pill-takers with diagnosis}}{\text{\% pill-takers in control group}}$

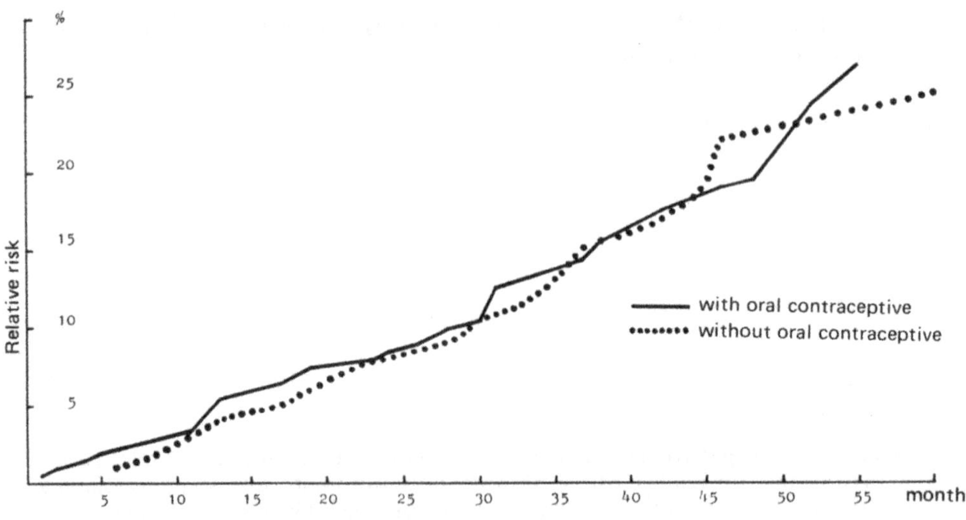

Fig. 1. Recurrence of fibrocystic disease in women using contraceptives (488 cases of fibrocystic disease diagnosed by biopsy in women from 15 to 55 years of age).

Table 2. Qualitative evaluation of normal appearing lobules in mammary parenchyme[a]

Lobular histology	On contraceptives	Not on contraceptives
End pieces		
No. of end pieces	114	100
Pycnosis of end pieces	93	100
Dense basal membrane	102	100
Secretion in lumen	116	100
Surrounding tissue		
Edema	95	100
Rich in cells	112	100
Fibrosis	99	100
No. of cases	n 56	n 26

[a] Women 18 to 35 years of age, with and without contraceptive use (index 100 for women not using contraceptives)

control group of women without manifest findings, a known percentage of whom had been taking contraceptives for an average of 30 months, with a population of women with manifest disorders as confirmed by biopsy. The risk of disease was found to be less than 1 (Table 1). An evaluation of first episodes and recurrences of fibrocystic disease over a 5-year period, has lent support to this conclusion (Prechtel 1978). In wo-

Table 3. Frequency of lobular secretion in normal appearing parenchyme[a]

Diagnosis	Secreting lobules	
	On contraceptives	Not on contraceptives
Fibroadenoma	2/24 = 8.3%	0/45 = 0
Fibrocystic disease I	0/69 = 0	7/139 = 5.0%
Fibrocystic disease II, III	0/19 = 0	6/53 = 11.3%
Carcinoma	3/30 = 10.0%	3/61 = 4.9%
Totals	5/142 = 3.5%	16/298 = 5.4%

[a] Biopsies diagnosed as fibroadenoma, fibrocystic disease, and carcinoma in woman 18 to 55 years of age

men whose biopsies showed fibrocystic disease recurrence did not depend on history of drug use (Fig. 1). One-fifteenth of our recurrent cases required a new biopsy to determine the course of the disease; here again, we found no one-sided distribution that would favor women who had been using contraceptives.

To our present knowledge, three endocrine hormones have an influence on the histomorphology of the mammary gland: the ovarian steroids estrogen and progesterone, and the pituitary hormone prolactin. In terms of their effect on the lobular system, we see three typical pictures:

1. Estrogen-dependent lobules show a surrounding tissue rich in mucopolysaccharides.
2. Gestagen-dependent lobules show a pronounced end-piece pattern.
3. Prolactin-stimulated lobules show acinar differentiation and active cellular secretion.

Since oral contraceptives have a direct effect on the body's hormone levels, we must ask whether they lead to structural changes in the mammary gland in terms of one or more of these patterns.

We looked at 82 cases of mild and uncomplicated fibrocystic disease in women 35 years of age and younger, evaluating the condition of their lobules in qualitative and semiquantitative terms. In the 26 women who had been using contraceptives for long periods of time, we found mean discrete changes, in that the number of end pieces had increased and the index of cellular pyknosis decreased. Furthermore, the surrounding stroma appeared richer in cells (Table 2). The pattern seen resembled that of the premenstrual phase (Meyer 1977).

In a further quantitative study we evaluated 440 cases of fibroadenoma, fibrocystic disease, and carcinoma in women 55 years of age and under (298 of whom had used and 142 not used contraceptives). Here we were interested in seeing whether the women on contraceptives showed a higher incidence of solitary lobular secretion or prolactin-stimulated lobules, and perhaps also galactorrhea. However, we found no indication that this type of pattern could be attributed to contraceptive use (Table 3).

What conclusions may be drawn from these findings? First, the parenchymal pattern in the end-piece region is generally better defined in women using contraceptives, due to an early and constant gestagen effect similar to that seen in the premenstrual phase. Yet there was no indication of secretory activity or increased epithelial proliferation, which may be significant. In no case, however, were we able to relate parenchymal pattern to hormonal medication with any reliability – the changes were too varied and discrete for that. Mammary gland parenchyme is not as acutely sensitive to steroids as endometrium. Thus in terms of the pathology of hormonal medication, the breast cannot be said to be the most unfavorably affected organ.

References

Fechner RF (1977) Influence of oral contraceptives on breast diseases. Cancer 39: 2764–2771

Hartmann E (1978) Mortalitätsrisiko – wegen oder nach der Pille? Aerztl Praxis 30: 543–548

Meyer JS (1977) Cell proliferation in normal human breast ducts, fibroadenomas, and other ductal hyperplasias measured by nuclear labelling with tritiated thymidine. Hum Pathol 8:67–81

Prechtel K (1978) Prospektive Untersuchungen über die Mastopathie und das Fibroadenom der Mamma. In: Grundmann E, Beck L (eds) Brustkrebs-Früherkennung. Fischer, Stuttgart New York, pp 109–113

Closing Remarks

In the last two days we have discussed and exposed ourselves to a wealth of information restricted to a relatively narrow field of study. As the theme of the meeting prescribed, the information presented dealt with topics of synergistic and antagonistic interactions of sex hormones, exposing their many opposite effects: there were those between estrogenic and progestagenic actions, those between clinical advantages and morphological changes, those between actions of single hormones causing serious endocrine imbalances, and actions of small doses of hormones adjusted to the patients' needs providing well-balanced effects. These differences particularly stress the need for us to build bridges of cooperation between our specialties, an effort we mentioned in the introduction. The exchange of information has taken place. Now we, the gynecologists, pathologists, biochemists, endocrinologists and anatomists must bridge the gulfs that separate our specialties to guarantee and promote the flow of information between us in the future.

We have been impressed again to see how synthetic and natural estrogens can induce tumors in animals, even carcinomas, and how diethylstilbestrol causes tumors in human beings. We have also been greatly interested to learn more about how gestagens are able to cause endometrial carcinomas to involute.

We have seen again that precancerous conditions of the portio develop earlier and more frequently in women taking oral contraceptives than in those without that therapy. These facts, supported by statistical studies, confirm what Professor Mohr presented before our working group ten years ago; his data at that time were regarded as controversial and consequently severely criticized.

We intentionally started the symposion by discussing the actions of estrogens separately from those of gestagens. That analysis should help us understand how combined hormonal preparations, especially the oral contraceptives, induce harmful side-effects, those we already recognize as well as those yet unknown. The facts presented confirmed once again that both types of hormones stimulate different structures in different ways. We learned that estrogens cause the squamous epithelium of the portio, the endometrium and columnar epithelium of the Fallopian tube to proliferate. In contrast, gestagens cause the endocervical mucosa to proliferate but the endometrium to atrophy. It was further shown that both hormones induce specific histological changes, for example, excessive gestagens cause microglandular hyperplasia of the endocervical mucosa as against the physiological cystic hyperplasia of pregnancy.

In our discussion, we were able to analyze the subject in another way, according to the organs involved. We then summarized the actions of hormones as a provisional working hypothesis. As we heard, hormones are administered for very diverse reasons. Our questions about their actions and criticism of their misuse assume all the more importance when one considers that it is young healthy women who take the oral contraceptives. In such patients we must regard even mild side-effects in a low percentage as serious. In contrast, in the group of patients with endometrial carcinomas treated with gestagens, we need only ask whether the risk of such therapy is less than that of no such therapy.

Even if we are unable to answer all final questions, at least we have seen more clearly where the problems lie, those concerned with the irresistible desire to administer hormone therapy for various reasons, and those concerned with deleterious effects, in part morphological. The pathologist should not become the bad conscience of the clinicians. Rather, he should serve as their critical companion, the aim of close and harmonious cooperation.

In concluding, I feel indebted to heartily thank you all for participating so enthusiastically in this symposion. My special thanks are directed to those speakers who shunned neither distance nor time to come here to present new concepts and rediscuss older ones, enabling us to exchange information at an international level.

This symposion, we hope, represents an auspicuous beginning, serving as a change and challenge to continue with our cooperation.

Gisela Dallenbach-Hellweg

Current Topics in Pathology

Continuation of
Ergebnisse der Pathologie

Editors:
E. Grundmann, W. H. Kirsten

Volume 62

Developmental Biology and Pathology

Editors: A. Gropp, K. Benirschke
With contributions by C. R. Austin, H. Beier, O. Bomsel-Helmreich, A. Boué, J. G. Boué, H. W. Denker, W. Engel, C. J. Epstein, V. H. Ferm, W. Franke, A. Gropp, M. Karkinen-Jääskeläinen, C. Lutwak-Mann, B. Putz, I. Saxén, L. Saxén, H. Spielmann, D. Szöllösi, U. Zimmermann

1976. 86 figures. IX, 218 pages
ISBN 3-540-07881-9

Volume 64

Pulmonary Hypertension Related to Aminorex Intake DNA Injuries, Their Repair, and Carcinogenesis Soft Tissue Tumors in the Rat. Visceral Candidosis

With contributions by K. Küchemann, J. L. Van Lancker, D. Mass, E. L. Quiroga, U. N. Riede, K. Salfelder, J. Schwarz, H. J. Steinhardt, C. Thomas, K. Ueda, S. Widgren

1977. 107 figures. VI, 228 pages
ISBN 3-540-08107-0

Volume 66

Perinatal Pathology

With contributions by M. Bibbo, C. Bron, W.-W. Höpker, J. P. Kraehenbuhl, B. Ohlendorf, L. Olding, S. Panem, B. Sandstedt, H. Soma, B. Sordat
Editors: E. Grundmann, W. H. Kirsten

1979. 88 figures, 34 tables. VI, 218 pages
ISBN 3-540-09207-2

Volume 67

Carcinogenesis

With contributions by P. Höhn, E. Kunze, K. Nomura, C. Witting, W. Schlake
Editor: E. Grundmann

1979. 112 figures, 21 tables. VI, 259 pages
ISBN 3-540-09344-3

Springer-Verlag
Berlin
Heidelberg
New York

Pathology of the Female Genital Tract

Editor: A. Blaustein
1977. 1206 figures, 39 four-color figures.
XX, 897 pages
ISBN 3-540-90180-9

Contents: Chapters on Anatomy, Histology, Cytology, Diseases, Benign and Malignant Tumors of Vulva, Vagina, Cervix, Endometrium, Myometrium, Uterus, Fallopian Tube, Ovary, Placenta, Membranes, Umbilical Cord, and Others.

This book was conceived as an all inclusive reference text for those individuals interested in gynecologic and obstetric pathology. All existing books were carefully viewed so that this edition could embody the salient and exceptional feature in the available literature as a comprehensive compendium on the subject.
Expanded information on the field of gynecologic pathology renders simple authorship obsolete. The topics have been carefully selected by the editor and assigned an outstanding group of contributors. The manuscript has been carefully edited so that the consistency and coordination prevails especially in the case of closely allied topics. A definite effort was made to provide quality illustrations which assume special importance in a pathologic treatise. Several topics are presented which currently do not exist in available texts.

Endocrine Treatment of Breast Cancer

A New Approach
Editors: B. Henningsen, F. Linder, C. Steichele
1980. 69 figures, 75 tables. Approx. 260 pages
(Recent Results in Cancer Research, Volume 71)
ISBN 3-540-09781-3

Contents: Improved Biochemical Characterization of Breast Cancer. – Antiestrogens in Experimental Breast Cancer. – Endocrine Treatment of Advanced Breast Cancer. – Anti-Oestrogens in the Treatment of Advanced Breast Cancer. – Combination Therapy of Advanced Breast Cancer. – Adjuvant Endocrine Therapy of Breast Cancer. – Principles of Clinical Trials. – Anti-Oestrogen Treatment in Breast Cancer. – A Comprehensive Review. – Subject Index.

This book summarizes the experience of leading, internationally recognized study groups in the field of hormone therapy for the treatment of patients with breast cancer. It presents and overall framework of possible methods of hormone therapy by including extirpative operative procedures.
Emphasis is placed on the redetermination of the indications for the individual methods of therapy. This redetermination has become essential since greater rates of remission have become possible by applying more appropriate therapy with fewer side effects by means of hormone receptor determination on the one hand and the development of substances with an antiestrogenic effect on the other hand.
Methods and problems of hormone receptor determination as well as experimental and clinical data pertaining to antiestrogen therapy are discussed in detail.
Furthermore, the discussion extends to the borderline between hormone therapy and chemotherapy and also covers adjuvant endocrine therapy of breast cancer.

H. Ludwig, H. Metzger

The Human Female Reproductive Tract

A Scanning Electron Microscopic Atlas
1976. 546 micrographs. XI, 247 pages
ISBN 3-540-07675-1

Contents: The Vagina. – The Ectocervix and Endocervix. – The Endometrium. – The Fallopian Tube. – The Ovary. – Gestational Metamorphosis of the Tissue Surface. – Metamorphosis of the Tissue Surface by Progestational Agents. – The Placenta. – The Membranes.

This atlas contains 546 scanning electron-microscopic images of the human female reproductive tract – the surface tissue of the ovaries, fallopian tubes, endometrium, cervix, portiones, and vagina in their various cyclical functional states, as well as during pregnancy and menopause. In addition, the internal surfaces of the placenta and fetal membrane are presented. Pathologic changes of these surfaces are considered insofar as they aid toward an understanding of the morphologic reactions to physiologic processes.
The illustrations are grouped in plates in such a way that x50 survey magnifications which reproduce the typical tissue architecture are supplemented by progressively higher, up to x 20000, detail magnifications.
The thorough text of the legends is not confined to their accompanying plates, but also contains references to data from physiology, pathophysiology, microscopic, anatomy, transmission electron microscopy, and histochemistry. The plates are introduced by a chapter on the methodology of preparation of specimens, and equipment, upon the use of which scanning electron microscopy largely depends. With its vivid three-dimensional plates, the atlas represents a substantial instructional vehicle and provides microanatomic images which the unpracticed can hardly derive from classic two-dimensional histologic sections.

CPSIA information can be obtained at www.ICGtesting.com
Printed in the USA
LVOW03s2043050814

397656LV00017B/472/P

9 783642 675706